PRE-SERVICE MATHEMATICS TEACHERS AS LESSON DESIGNERS

This book addresses how pre-service mathematics teachers (PMTs) can be better prepared to design and teach lessons that engage students in reasoning and proving (R&P), a core yet persistently challenging aspect of mathematics education. Building on the European MaTeK project, the book presents a multi-phase intervention developed collaboratively by five universities to enhance PMTs' design capacity in teaching R&P across diverse curricular contexts.

Grounded in current international research, the book combines theoretical and empirical perspectives on R&P, lesson design, and teacher design capacity. It analyzes PMTs' evolving understanding of reasoning, their task and lesson designs, and their reflections on classroom implementation. The findings highlight the transformative potential of international collaboration and structured feedback in preparing teachers for contemporary classrooms.

Combining conceptual clarity with practical examples, the book offers valuable insights for mathematics teacher educators, researchers, and curriculum developers. It contributes to ongoing discussions on how to support teachers in creating learning environments where mathematical reasoning and proof are integral to student learning.

Mária Slavíčková is Associate Professor of Didactics of Mathematics at Comenius University in Bratislava, Slovakia. Her research focuses on preparing future mathematics teachers, with emphasis on reasoning, proof, lesson design, and integrating technology in teaching.

Nad'a Vondrová is Full Professor of Mathematics Education at Charles University, Czech Republic. Her research includes teacher knowledge, word problems, and integrating mathematical and language literacy.

Birgit Pepin is Emeritus Professor of Mathematics Education, formerly holding professorships in Norway and the Netherlands. Her research focuses on mathematics curriculum resources, digital tools, and innovative teaching environments.

Iveta Kohanová is Associate Professor of Mathematics Education at the Norwegian University of Science and Technology (NTNU). Her work focuses on teacher development, assessment, and the use of resources in mathematics education.

Benedetto Di Paola is Associate Professor of Mathematics Education at the University of Palermo (UniPa), Italy. He is Coordinator of the Mathematics Education Research Group (GRIM) and Vice-President of Commission for the Study and Improvement of Mathematics Teaching (CIEAEM), with research on teaching practices in mathematics.

Jakub Michal is Assistant Professor at the Faculty of Education, Charles University, Czech Republic. His research focuses on reasoning and proving in school mathematics, teachers' conceptions of proof, and instructional resources.

Işıl İşler-Baykal is Assistant Professor of Mathematics Education at Middle East Technical University (METU), Turkey. Her research focuses on algebraic thinking, reasoning, proof, and teacher education.

Didem Akyüz is Professor of Mathematics and Science Education at Middle East Technical University (METU), Turkey. Her research focuses on design-based research, teacher education, and technological pedagogical content knowledge.

New Perspectives on Research in Mathematics Education – ERME series

The European Society for Research in Mathematics Education (ERME) represents a vibrant and growing community of approximately 2500 researchers from across Europe and beyond. United by a shared commitment to understanding and improving the teaching and learning of mathematics, ERME fosters collaboration across borders, disciplines, and educational contexts.

The *ERME Book Series*, published by Routledge, is a key vehicle for documenting and disseminating the substantial and diverse body of research emerging from this community. Volumes in the series include scholarly monographs and edited collections that grow out of sustained collaborations among European researchers in mathematics education. While firmly rooted in the European tradition, the series is international in scope – written by and for researchers not only within Europe, but also around the world.

A hallmark of the series is its commitment to rigorous academic standards. Each volume undergoes a thorough and constructive peer review process, overseen by an international advisory board that ensures both scholarly excellence and global relevance. This process helps maintain the quality and integrity of the series, while encouraging innovative and critical scholarship.

The contributions featured in the *ERME Book Series* aim to deepen our understanding of mathematics education across a wide range of settings: from schools and universities to informal learning environments. A particular emphasis is placed on examining the institutional, societal, and cultural contexts that shape how mathematics is taught, learned, and researched. The series embraces a plurality of perspectives, theoretical frameworks, and methodological approaches, reflecting the richness and diversity of the field.

By fostering wide authorship and encouraging dialogue across contexts, the *ERME Book Series* contributes to a shared, yet multifaceted, vision of mathematics education research – one that continues to evolve in response to the needs and challenges of our time.

Volumes of the series:

(2) Working with the Anthropological Theory of the Didactic in Mathematics Education
A comprehensive casebook
Edited by Marianna Bosch, Yves Chevallard, F. Javier García, and John Monaghan

(3) Professional Development and Knowledge of Mathematics Teachers
Edited by Stefan Zehetmeier, Despina Potari, and Miguel Ribeiro

(4) Classroom Research on Mathematics and Language
Seeing Learners and Teachers Differently
Edited by Núria Planas, Candia Morgan, and Marcus Schütte

(5) Research and Development in University Mathematics Education
Overview Produced by the International Network for Research in Didactics of University Mathematics
Edited by Viviane Durand-Guerrier, Reinhard Hochmuth, Elena Nardi, and Carl Winsløw

(6) Mathematics Education in the Digital Age
Learning, Practice and Theory
Edited by Alison Clark-Wilson, Ana Donevska-Todorova, Eleonora Faggiano, Jana Trgalova, and Hans Georg Weigand

(7) Digital Technology and Artificial Intelligence in Mathematics Education
Edited by Eirini Geraniou, Cosette Crisan, and Manolis Mavrikis

(8) Pre-Service Mathematics Teachers as Lesson Designers
Mária Slavíčková, Nad'a Vondrová, Birgit Pepin, Iveta Kohanová, Benedetto Di Paola, Jakub Michal, Işıl İşler-Baykal, and Didem Akyüz

For more information about this series, please visit: https://www.routledge.com/European-Research-in-Mathematics-Education/book-series/ERME

PRE-SERVICE MATHEMATICS TEACHERS AS LESSON DESIGNERS

Mária Slavíčková, Naďa Vondrová, Birgit Pepin, Iveta Kohanová, Benedetto Di Paola, Jakub Michal, Işıl İşler-Baykal, and Didem Akyüz

Routledge
Taylor & Francis Group
LONDON AND NEW YORK

Cover Image: © Getty Images

First published 2026
by Routledge
4 Park Square, Milton Park, Abingdon, Oxon OX14 4RN

and by Routledge
605 Third Avenue, New York, NY 10158

Routledge is an imprint of the Taylor & Francis Group, an informa business

British Library Cataloguing-in-Publication Data
A catalogue record for this book is available from the British Library

ISBN: 978-1-032-87675-7 (hbk)
ISBN: 978-1-032-87674-0 (pbk)
ISBN: 978-1-003-53387-0 (ebk)

DOI: 10.4324/9781003533870

Typeset in Optima LT
by SPi Technologies India Pvt Ltd (Straive)

CONTENTS

ABOUT THE AUTHORS

Didem Akyüz, Professor of Mathematics and Science Education at METU, completed her doctoral studies at the University of Central Florida in the Department of Mathematics Education. During her doctoral studies, she investigated effective teaching approaches of an expert mathematics teacher regarding planning and classroom practices. She also served as a researcher on a STEM project combining technology, mathematics education, and computer engineering. She has been a faculty member in the Department of Mathematics and Science Education at METU since 2012. Her research focuses primarily on design-based research, teacher education, the practices of teachers and pre-service teachers, and technological pedagogical content knowledge.

Benedetto Di Paola, PhD, is Associate Professor of Mathematics Education at the Department of Mathematics and Computer Science at the University of Palermo (UniPa), Italy. He is the current Coordinator of the *Mathematics Education Research Group* (GRIM) and Director of the qualifying courses for teaching mathematics at all school levels at UniPa. He is also a member of the *Teaching and Learning Centre for Innovation and Improvement in University Teaching (TLC-CIMDU)*. Professor Di Paola is Vice-President of the *Commission for the Study and Improvement of Mathematics Teaching* (CIEAEM) and the *Association for the Research of Mathematics Teaching for All Students* (M-CTRAS). As a teacher educator, Professor Di Paola teaches numerous courses on the didactic and history of mathematics in primary and secondary education university programmes. He is also involved in several PhD programmes in mathematics education, supervising PhD students and teaching doctoral courses in this research area. Di Paola is the author of numerous publications; his main research interests focus on studying teaching practices in mathematics from kindergarten to university from historical, social, and cultural perspectives. He sits on the editorial boards of several international journals and books on mathematics education and is currently Editor-in-Chief of the international journal *Quaderni di Ricerca in Didattica*.

Iveta Kohanová is a researcher and teacher educator in mathematics education at the Department of Teacher Education at the Norwegian University of Science and Technology (NTNU), where she has worked since 2017. She is involved mainly in lower secondary mathematics teacher education, and her research focuses on the use of various resources for and in mathematics education, mathematics teachers' professional development, and assessment in mathematics education. She holds master's degrees in mathematics and computer science education and a PhD in mathematics education from Comenius University in Bratislava, Slovakia. Before moving to Norway, she taught mathematics at a grammar school in Bratislava for fifteen years, an experience that continues to inform her research and her work with pre-service teachers.

Işıl İşler-Baykal is Assistant Professor of Mathematics Education at the Middle East Technical University (METU) in Ankara, Turkey. She completed her PhD at the University of Wisconsin-Madison, where she worked on a series of projects, Project LEAP (Learning through an Early Algebra Progression). She continues to work on algebraic thinking, reasoning, and proof in school mathematics and teacher education. She served as a member of the topic group Teaching and Learning of Early Algebra at the 15th International Congress on Mathematical Education (ICME). She teaches several courses at the undergraduate and graduate levels, including methods of teaching mathematics and theoretical perspectives in mathematics education. She has published in several journals, including Journal for Research in Mathematics Education, Educational Studies in Mathematics, and Journal of Mathematics Teacher Education.

Jakub Michal is Assistant Professor at the Faculty of Education, Charles University, where he teaches university mathematics courses and works in the field of mathematics education. He completed his PhD with a dissertation that explored how pre-service teachers can be supported in developing and teaching mathematical reasoning and argumentation. His research focuses on reasoning and proving in school mathematics, teachers' conceptions and beliefs about proof, and the use of instructional resources in mathematics teaching. He has co-authored studies published in several conference proceedings, as well as articles examining the role of digital technologies in mathematics education. In addition, he specializes in analogue mathematical instruments – such as planimeters and other traditional tools – that support deeper understanding of geometric and algebraic ideas. He regularly participates in international research conferences as well as practice-oriented conferences for in-service mathematics teachers, and he has also been involved in organizing several local professional events and conferences. At the university, he teaches courses in linear algebra, number theory, polynomial algebra, and other mathematics-related subjects.

Birgit Pepin taught mathematics in schools in England before starting her PhD in mathematics education, a comparative study of mathematics teachers' work in England, France, and Germany, which related to her background (German) and the countries in which she lived and worked (UK, France). Subsequently, she worked in UK higher education institutions (Open University, Oxford Brookes University, and Manchester University). Subsequently, she took up a full professorship in Norway (2009–2014), followed by the Netherlands (2015), where she became emeritus professor on retirement in 2023. She has gained research funds from prestigious research funding institutions (e.g., ESRC and EU) and has served on many research councils and advisory boards. Moreover, she has published, and still publishes, widely in the field of (e-) curriculum resources, and teacher and student appropriation of such resources for mathematics teaching and learning. Her work focuses on mathematics curriculum resources, in particular digital and AI-related resources, and teacher pedagogy and student learning in innovative environments. Her professional interests include the improvement of mathematics teaching and of AI- and digital-resources-supported education more broadly; the design and integration of innovative educative tools and resources; and the education and professional development of teachers.

Mária Slavíčková is Associate Professor of Didactics of Mathematics at Comenius University in Bratislava, Slovakia. Her research focuses on the preparation of future mathematics teachers, with particular emphasis on reasoning and proof, lesson design, and the integration of technology in teaching. She has led innovative projects in mathematics teacher education, including the European Horizon 2020 MaTeK project (2021–2023), which strengthened pre-service teachers' reasoning and proof skills across five European universities. Mária has developed online and hybrid courses, co-authored textbooks, and supervised several bachelor's, master's, and PhD theses. She is actively engaged in curriculum development, national and international conferences, and scholarly publishing, including serving on editorial boards and conference committees. Her work bridges research and practice, providing evidence-based strategies to enhance teacher design capacity and improve mathematics instruction in diverse contexts.

Nad'a Vondrová, PhD, is a researcher and teacher educator in mathematics education at the Faculty of Education, Charles University, Prague, where she is currently a Full Professor. She holds qualifications in mathematics education, English teaching, and mathematics and geography teaching, and completed her PhD in mathematics education. She currently serves as Head of the Department of Mathematics and Mathematical Education and as Chair of the Board of Doctoral Studies in Mathematics Education. She also participates in university-wide coordination of doctoral studies in subject didactics and contributes to national professional organizations. Professor Vondrová has been involved in a range of research projects related to mathematics education, including studies on teacher knowledge, word problems, and the integration of mathematical and language literacy. Her work spans both national grants and international collaborations such as Horizon 2020. She has also taught mathematics at the primary and secondary levels and has experience with international teaching contexts.

ACKNOWLEDGEMENTS

This book is an outcome of the H2020 project MaTeK, no. 951882. We extend our sincere thanks to Tünde Kozánek Kiss for her valuable support with data processing when preparing this book. We are also grateful to all participating PMTs for their willingness to share their work, reflections, and feedback.

1
INTRODUCTION

Reasoning and proving (R&P) are central to learning mathematics with understanding (e.g., Hanna and de Villiers, 2012; Schoenfeld, 2011) and are seen as essential to learning mathematics in secondary school. The importance of R&P in school education is also reflected in policy documents such as the PISA 2022 mathematics framework (OECD, 2022), the new National Curriculum in Slovakia (MinEdu, 2023), Norway (Ministry of Education and Research, 2019), the Czech Republic (MŠMT, 2023), Italy (MIUR, 2010, 2025), Türkiye (MoNE, 2024a, 2024b), or in the United States (NCTM, 2009). Indeed, mathematicians often see reasoning as the core activity in mathematics education. At the same time, recent international literature reviews (e.g., Mariotti et al., 2018; Stylianides et al., 2017) have shown that proof has been challenging to learn and teach, and moreover to integrate R&P as an essential component of mathematics education. Studies report that teachers tend to focus on procedural skills that are easier to teach and assess (Harel & Rabin, 2010).

Hanna and Barbeau (2010) emphasise that in mathematics classrooms, proofs can be bearers of mathematical knowledge by conveying methods and strategies, and explaining why mathematical results are true, which Buchbinder and McCrone (2022) call *teaching mathematics* via *reasoning and proving* (p. 4). Stylianides et al. (2017) report on a vast body of knowledge on pupils' difficulties with proving, and teachers' challenges in teaching it; several theoretical frameworks explain the origins of student challenges with proving. However, there is little knowledge on how teachers can successfully create learning environments that support student engagement with R&P (Stylianides et al., 2017). As importantly, there is little theoretical or practical knowledge on preparing teachers to design (and enact) this type of mathematics teaching.

DOI: 10.4324/9781003533870-1

This is the area that we have addressed in the project and that is the basis for this book.

Teachers are seen as playing a key role in shaping and changing the culture of mathematics classrooms (e.g., Nardi & Knuth, 2017), and the role of the teacher is shaped by the goal of facilitating understanding of R&P. However, mathematics teachers themselves often lack knowledge and experience about different ways of R&P, about preparing a lesson focused on R&P, and about teaching R&P itself. Hence, they do not support pupils to learn R&P (Ko, 2010). Moreover, teachers might not have a strong sense of what R&P may look like in secondary mathematics besides what they find in school textbooks.

To address the need to better prepare teachers for the challenges associated with teaching R&P, we developed and systematically studied an intervention specifically designed for this purpose. The work was conducted within the framework of the European Union-funded MaTeK project (Enhancement of Research Excellence in Mathematics Teacher Knowledge), which aimed to strengthen the research and educational capacities of the Department of Mathematics Education at Comenius University in Bratislava. In this context, our goal was to enhance pre-service mathematics teachers' (PMTs) design capacity in Slovakia, with a particular focus on R&P. To achieve this, the Slovak research team collaborated with partners from four European institutions: Charles University (Czech Republic), the University of Palermo (Italy), the Norwegian University of Science and Technology (Norway), and the Middle East Technical University (Türkiye), who contributed their complementary expertise in mathematics education and teacher development.

Together, the international consortium co-designed and implemented a multi-phase intervention embedded in PMT education. These guided the PMTs through cycles of

(1) learning about different ways of R&P,
(2) preparing mathematical R&P tasks and lesson plans (or for a learning trajectory) for teaching in secondary classrooms (that integrate these R&P aspects with the curriculum of their country),
(3) reflecting on the lesson plans produced by the different groups in their university,
(4) implementing one (commonly agreed and chosen) lesson plan in a secondary classroom, and
(5) reflecting on the implemented lesson in order to amend the lesson plan.

Based on a presentation of each country's mathematics curriculum, the final lesson plans of each country were then (6) presented and discussed (by PMTs and teacher educators involved in the project) in an international meeting, followed by (7) the development (by internationally mixed groups) of a new set of lesson plans on a set curriculum topic area. These new lesson plans were subsequently presented and discussed with the meeting participants in the plenary meeting.

Hence, this project aimed to design and evaluate an intervention to advance PMTs' lesson design capacity (see Pepin et al., 2017; Trgalová & Tabach, 2024) for preparing to teach R&P, and to investigate the challenges and achievements that PMTs had when learning to design lessons to teach R&P.

This book presents our theoretical frames (e.g., R&P, ways of reasoning, lesson design, teacher design capacity). Subsequently, we explain how this guided the design of the intervention, in particular the lesson planning component. The results are then presented, including analyses of (1) the lesson plans and their development over time, focusing on ways of using different ways of reasoning; and (2) PMTs interviews with respect to their developing ideas of how to view R&P in their mathematics teaching and pupil understanding. We end with insights and conclusive statements we developed from the intervention.

2

THEORETICAL BACKGROUND

Becoming a teacher requires developing substantial knowledge of the content, the curriculum and curriculum resources, the instructional approaches, in addition to pupils' interests and beliefs, to be able to teach mathematics effectively (e.g., Chen & Zhang, 2019). Lim et al. (2018) reported that PMTs find it difficult to recognise the mathematical idea within the task and transfer it to a lesson plan. Chen and Zhang (2019) also discovered that PMTs lacked knowledge of how to modify planning aspects to meet the needs of their learning groups. A lesson plan can support PMTs' developing ideas and offer support to enhance and deepen their knowledge. It will likely help them think about the lesson's main aims, how they would want their students to learn, and how they can assess student learning experiences.

Over recent decades, the central role of reasoning and proving (R&P) in pupils' engagement with and learning of mathematics has been widely recognised. Nevertheless, there remain relatively few systematic attempts to design and structure mathematical experiences that place R&P at the core of instruction (e.g., Stylianides et al., 2024; Valenta et al., 2024; Weingarden & Buchbinder, 2023). This continuing imbalance between the acknowledged importance of R&P and its limited classroom realisation underscores the need to better understand how teachers can be supported in planning and enacting lessons that cultivate students' R&P practices.

Therefore, we examined how PMTs can be supported in developing the design capacity, implement, and reflect on lessons that foster R&P. First, we explain key concepts of our study, such as lesson design capacity, lesson planning and task (re)design, and teacher reflection and noticing. Second, we elaborate on perceptions and roles of mathematical reasoning, on PMTs'

DOI: 10.4324/9781003533870-2

perspectives about R&P, and its teaching and learning. In the third section, we provide the research questions of the study.

Our work is grounded in the Documentational Approach to Didactics (DAD; Trouche et al., 2019), which provides a conceptual framework for analysing teachers' interactions with resources and the development of their professional knowledge through documentational processes.

2.1 Lesson design capacity and lesson planning

In this section, we explain and review the literature in terms of three key concepts of this study: lesson design capacity, lesson planning and task design, and teacher noticing and reflection.

2.1.1 Lesson design capacity

Planning a mathematics lesson is a complex process and, simultaneously, a core teaching task. Lesson planning is crucial for developing teachers' didactical fluency (Pepin, 2015) and good mathematics instruction (e.g., Chen & Zhang, 2019; Yunianto et al., 2021). It is often considered an important part of teachers' pedagogical knowledge and practice (Ball & Forzani, 2009; Hogan et al., 2003), and in that sense, lesson plans have been used productively in models of teacher professional learning. It is assumed that the quality of lesson planning contributes to the effectiveness of mathematics instruction and, finally, to student learning (e.g., Fitriati et al., 2023). Given this significance, supporting PMTs in developing robust lesson planning skills is essential in initial teacher education, as it plays a critical role in their growth towards becoming effective and reflective practitioners (Pepin, 2015).

In the mathematics education research literature, the notion of teacher design and teacher design capacity is relatively new. Despite its novelty, its conceptualisation has evolved over the past decade, mainly due to (i) the introduction and appropriation of digital resources in mathematics education (e.g., Pepin et al., 2017) and (ii) the emphasis on teacher collaborative work.

In 2017, Pepin and colleagues argued that teaching can be considered design. This was in line with earlier research by Brown (2009), who explains that "teachers must perceive and interpret existing resources, evaluate the constraints of the classroom setting, balance trade-offs, and devise strategies – all in the pursuit of their instructional goals" (p. 18). He claims that the interpretation of teaching as design, and the notion of teachers as designers, fits with a range of cognitive theories that

> emphasise the vital partnership that exists between individuals and the tools they use to accomplish their goals. [...] And it is not just the

> capacities of individuals that dictate human accomplishment, but also the affordances of the artefacts they use
>
> *(p. 19).*

Hence, Brown (2009) and Pepin et al. (2017) regarded this relationship as an interrelationship: the activity of "designing" is not only dependent on the teacher's competence, but it is an interrelationship between the teacher/s and the (curriculum) material/s, the teacher-tool relationship, that is at play here. Hence, the affordances of the curriculum materials influence this relationship as much as the teacher's competence and knowledge.

As mentioned elsewhere (Pepin et al., 2017), the notion of "design" is understood (a) as the practice of designing for teaching, as in lesson preparation (i.e., design before enactment), (b) as well as in teaching, what is often labelled as "design-in-use" that happens during enactment of the resources/materials (e.g., Pepin et al., 2013). However, any understanding of teacher-as-designer must include a conscious/deliberate act of designing, creating "something new" (e.g., combining existing and novel elements) to reach a certain (didactical) aim. This must be true for both (a) and (b). Hence, any act of designing would include the teacher being explicit (and conscious), in particular about the following:

1 why they are designing in this particular way,
2 which aims/goals they have in mind,
3 who is their audience,
4 which resources and tools they are using for the design,
5 how they plan to evaluate the designed lesson plan.

Moving on to "design capacity", "pedagogical design capacity" was a term coined by Brown (2009) and "teacher design capacity" by Pepin and colleagues (Pepin et al., 2017). Brown describes pedagogical design capacity as the capacity to utilise and transform existing curricular resources effectively and to design/create new materials for effective mathematics instruction. He identified three types of curriculum use: offloading, adapting, and improvising, and proposed a framework for describing teachers' capacity to design with curriculum materials (Design Capacity for Enactment Framework), which was later further developed by Remillard and Heck (2014). Pepin et al. (2017) describe *teacher design capacity* in a more goal-oriented way by emphasising that teacher design capacity should include the following components:

- a goal, an orientation, or point/s of reference for the design,
- a set of design principles, which must be firm but flexible,
- a "reflection-in-action" type of implicit understandings and realisations (most visible in action, that is during teaching as "design-in-use").

However, later Pepin et al. (2019) claim that the earlier definition provided "an ambitious, idealised image of teacher design capacity" (p. 125). They propose a new model for teacher design activities, in which teachers' *modes of design* are stretched by two dimensions: *the teaming dimension* (from working alone to working in a collective) and *the use dimension* (from own use to generic use).

It should be noted that both conceptualisations/definitions are theoretically underpinned by socio-cultural theory, in particular the literature on the mediating role of artefacts (e.g., Wartofsky, 1973); and it is clear that design/ing is a purposeful activity and should not be interpreted as a "subconscious" act.

Rather than focusing on the individual teacher, newer approaches to teacher design emphasise the importance of collaborative work and the new opportunities that Design Research offers (Pepin et al., 2017). These approaches often build on a broad conceptualisation of resources as "that" a teacher uses to resource their classroom teaching, including material and non-material resources, such as their knowledge or talk with colleagues (Adler, 2000). The general understandings of what constitutes a resource and what their roles are have thus been broadened within the research field (Trouche et al., 2019).

In designing our intervention to prepare PMTs for R&P lessons, we incorporated insights from curriculum development research, particularly the frameworks and tools described by van den Akker (2004), to clarify essential considerations in lesson development. For a comprehensive description of our approach to lesson planning, please refer to the methods section under Chapter 4.

Additionally, recent research highlights evolving trends in the concept of "teacher design". Psycharis et al. (2023) identify three key shifts in this area: (1) the role of the individual teacher has changed, with greater emphasis on collaborative design processes; (2) teacher collaboration has become more central, reinforcing the value of shared expertise; and (3) curriculum materials themselves have assumed a more dynamic role in shaping instruction, to the extent that recent materials are even viewed as having "agency" (Pepin et al., 2023). These shifts underscore the importance of equipping teachers to effectively design and adapt curriculum in a collaborative, responsive manner that aligns with the demands of contemporary educational practice.

2.1.2 Lesson planning and task (re)design

Effective *lesson planning* is at the heart of teachers' work. It requires thinking about "what to teach, how to represent it, how to question students about it and how to deal with problems or misunderstanding" (Shulman, 1986, p. 8). Planning lessons can improve their pedagogical content knowledge (Cevikbas et al., 2024) and impact the quality of their teaching. For example, Orr and Bieda (2023) found evidence that the PMTs who included significantly lower academically rigorous questions in their plans also used fewer such questions in their lessons. Both teachers and future teachers see planning lessons as a practical and authentic activity that is worthwhile to spend their time on (Watson & Sullivan, 2008), which makes it a beneficial part of teacher education programmes.

Moreover, collaborative activities, including collaborative design practices, are considered beneficial for teacher learning (Backfisch et al., 2023). Collaborative lesson planning provides space for teachers to develop "a shared meaning on the selection and use of curriculum resources in lesson plans as a result of the negotiation and a common decision process of the basis for individual decisions" (Ulusoy & Incikabi, 2023, p. 561). The cooperative planning process might encourage PMTs to think through their plan more thoroughly, "recognising possible hurdles to overcome and factors to be taken into account to a greater extent than when planning a lesson individually" (Vondrová et al., 2020, p. 34).

The key role in planning and enacting mathematics lessons is played by *mathematics tasks*. The act of analysing and modifying tasks draws teachers' attention to student learning (Watson & Sullivan, 2008). Task design is a complex and multifaceted process that requires careful attention to several key factors, including what to design, which tools to use, and under what conditions tasks should be implemented. As Thanheiser et al. (2015) highlighted, task design can involve creating individual tasks, groups of tasks, or even task sequences that align with specific learning goals. Moreover, the tools used in task design play a significant role in shaping the effectiveness of tasks. Hansen et al. (2016) emphasise the importance of manipulatives, while Tempier (2015) focuses on the affordances and constraints of different tools for selected topic areas. Cognitive tools, as discussed by Thanheiser et al. (2015, e.g., "reflective analysis"), also provide unique support for task design by helping to structure and deepen students' engagement with mathematical concepts.

Understanding the conditions under which tasks are implemented is crucial for their success. Johnson et al. (2015) highlight the potential tensions between task designers and practitioners, emphasising the need to address these challenges to ensure tasks are both practical and effective in classroom settings.

For PMTs, it is vital to be able to select mathematical tasks (e.g., from textbooks), to appropriate them for their lesson/s, to design instructional units (including the academic year's curriculum), a series of lessons, or even a single lesson. However, pre-service teachers find it difficult to produce their own tasks and tend to rely on textbooks or Internet resources (Ulusoy & İncikabi, 2023). Thus, redesigning a task might be more suitable than designing a task.

Jones and Pepin (2016) argue that involving teachers in task design enhances the relevance of tasks, ensuring that tools and conditions are carefully considered. This process promotes both professional learning and innovative research, fostering partnerships between teachers and researchers that refine task design principles and integrate them effectively with pedagogical approaches. Engaging PMTs in task redesign is integral to their preparation, as it encourages them to critically evaluate mathematical content, instructional strategies, and the diverse needs of learners. Through this iterative process, PMTs develop a deeper understanding of how tasks might influence student learning and how they can be modified to achieve specific instructional goals. Wenger (1998) identifies communities of practice, supported by shared knowledge resources like journals, as vital for sustained professional growth. Goldsmith et al. (2014) add that access to research enhances teacher learning communities, providing a common foundation for collaboration, reflection, and the development of best practices.

In line with these views, Doerr and Lesh (2003) emphasise that task redesign allows PMTs to engage actively with tasks, refining both their problem-solving and teaching strategies. This modelling process strengthens PMTs' sense of agency, making them contributors to task design rather than mere users. Liljedahl (2015) further supports this by noting that task redesign builds teacher capacity, encouraging reflection on how tasks align with curricula and learning objectives. Jaworski (2006) highlights the role of critical inquiry in task redesign, noting that collaborative discussions and iterative refinement allow PMTs to bridge theoretical concepts with practical applications, ensuring tasks are not only adapted but thoughtfully modified to meet the needs of students.

Task redesign also enhances PMTs' pedagogical content knowledge (Shulman, 1986). Ball et al. (2008) further reinforce this by underscoring how task redesign contributes to the development of "specialised content knowledge", helping PMTs balance curriculum demands with the diverse needs of students. Smith and Stein (2011) stress the importance of task selection and modification for fostering productive mathematics discussions. By redesigning tasks, PMTs learn to scaffold discussions, create tasks that stimulate meaningful mathematical discourse, and adapt tasks to meet diverse student needs. Zazkis and Leikin (2010) echo this view, noting that task redesign allows PMTs to balance mathematical rigour with accessibility, ensuring tasks promote equitable learning opportunities for all students. Furthermore, Watson and Ohtani (2015) highlight the significance of collaboration in task redesign, where partnerships with experienced teachers and researchers provide valuable feedback and diverse perspectives, helping PMTs refine tasks based on classroom experiences.

Watson and Ohtani (2015) highlight that incorporating digital tools into task design provides added affordances for developing dynamic learning environments. Building on this, Geiger et al. (2016) emphasise that partnerships involving PMTs, teacher educators, and researchers create valuable opportunities for PMTs to engage with task design principles and connect them with pedagogical strategies. Kieran et al. (2015) further argue that integrating task design into teacher education programmes is essential for equipping PMTs with the competencies needed to critically assess, adapt, and implement tasks in their future teaching.

In conclusion, task redesign is a critical component of PMTs' professional preparation. By engaging in this process, PMTs gain deep insights into task design, which fosters their ability to create meaningful and engaging learning experiences. Task redesign not only prepares PMTs for the practical realities of teaching but also empowers them to become reflective, adaptive, and innovative teachers capable of contributing to the ongoing improvement of mathematics education.

2.1.3 Teacher noticing and reflection

In the classroom, teachers must make continuous, in-the-moment decisions while navigating a complex instructional environment where multiple events occur simultaneously (van Es & Sherin, 2002). To manage this complexity, teachers rely on their capacity to notice salient aspects of students' thinking and learning (Dietiker et al., 2023) and to engage in reflective processes that support the interpretation of these interactions (Clarà et al., 2019). While noticing enables identifying and prioritising critical information, reflection facilitates the refinement of instructional responses, thereby enhancing students' mathematical understanding.

The interplay between noticing and reflection is central to effective teaching. Whereas noticing allows teachers to attend to significant moments as they unfold, reflection deepens their understanding and informs future pedagogical choices. Ward and McCotter (2004) identified three recurrent characteristics of frameworks for teacher reflection: it is situated in practice, cyclical in nature (often prompted by journaling or narrative writing), and shaped by multiple perspectives (e.g., peer dialogue). Mason (2011) underscores the importance of noticing as a precursor to reflection, arguing that "intentional reflection and reconstruction enhance the possibility of being sufficiently awake at some future moment to be able to respond freshly rather than to react habitually" (p. 41). In his view, reflection is not only a product of noticing but also a condition for more responsive noticing in future classroom situations. Similarly, van Es and Sherin (2008) highlight that "learning to notice is one important dimension within the process of reflection" (p. 247), further illustrating the reciprocal relationship between the two.

This dynamic relationship is further elaborated in the revised framework of teacher noticing (van Es & Sherin, 2024), which conceptualises noticing through three interrelated dimensions: attending, interpreting, and shaping. These dimensions are particularly relevant for PMTs, as they describe the processes through which teachers identify meaningful classroom events, interpret them using evolving pedagogical and mathematical knowledge, and construct instructional moves that respond to and build on student thinking. Importantly, the framework promotes an inquiry-oriented stance, encouraging PMTs to engage more critically and deeply with the classroom events they observe.

Noticing is thus both a cognitive and professional practice central to mathematics instruction. It supports teachers' own mathematical thinking (Sherin et al., 2010) and plays a critical role in high-quality instruction by enabling teachers to distinguish pedagogically significant moments from less relevant distractions (Fernández et al., 2020; Jacobs & Spangler, 2017). Recent research has further emphasised the relationship between PMTs' pedagogical content knowledge (Shulman, 1986) and their ability to notice students' mathematical thinking. Noticing is a mediating mechanism that links pedagogical content knowledge with teachers' responsiveness to students' ideas and their overall mathematical knowledge for teaching (Otten et al., 2019; Thomas et al., 2017). As such, noticing is widely acknowledged as a key professional competence in developing effective teaching practice (Sherin et al., 2010).

Despite the importance of noticing, opportunities for PMTs to develop this skill in live classrooms are often limited. In response, teacher education programmes frequently incorporate video-based resources to approximate real-world teaching scenarios (Mitchell & Marin, 2015; Roller, 2016; Star & Strickland, 2008). These video-recorded lessons allow PMTs to pause at key moments, revisit and analyse classroom interactions, and engage in deeper reflection (Santagata & Guarino, 2011). Furthermore, written reflections based on such video excerpts or other instructional artefacts offer additional opportunities for PMTs to develop professional noticing and articulate their interpretations (Sánchez-Matamoros et al., 2019).

Our study situates itself within this theoretical landscape by examining PMTs' capacity to plan and reflect on lessons. From this perspective, teaching is conceptualised not as the straightforward enactment of curricular materials, but as a complex, situated activity shaped by a teacher's subject-matter knowledge and pedagogical content knowledge (Ball et al., 2008), beliefs about teaching and learning (e.g., Philipp, 2007; Thompson, 1992), prior educational experiences, and professional dispositions (Remillard, 2005). This view aligns with the instrumental and participatory approaches to teaching resources, which highlight the evolving, negotiated relationship between teachers and the materials they use (Trouche et al., 2023).

2.2 Reasoning and proving in mathematics education

Teaching mathematics is intellectually and emotionally demanding (e.g., Stylianides & Stylianides, 2013). In teacher education, PMTs are helped to develop their skills for preparing to teach and for teaching mathematics in class. Stylianides et al. (2017) found a wealth of research on student struggles with proof and frameworks explaining these difficulties. Yet, they note a knowledge gap in research on supporting teachers in engaging with various forms of R&P to design effective lesson plans and learning trajectories. This gap extends to understanding how best to prepare teachers to enact R&P-centred teaching and which interventions might help PMTs develop skills in this field.

The research literature in mathematics education has broadly recognised the critical role of R&P in mathematics education (e.g., Hanna & de Villiers, 2012; Harel & Sowder, 2007). At the same time, policy documents and national standards in countries worldwide stress the importance of teaching and learning mathematics with understanding, including the reasoning behind mathematical procedures and theorems, being able to prove results in algebra and geometry, justify their thinking, and produce valid arguments for claims (e.g., NCTM, 2009). An underlying idea behind the emphasis on R&P is that they are key processes by which mathematical knowledge is constructed and validated; hence, they are important vehicles for learning and understanding mathematics (Weingarden & Buchbinder, 2023). Hanna and Barbeau (2010) indicate that in mathematics classrooms, proofs can be bearers of mathematical knowledge by conveying methods and strategies, and explaining why mathematical results are true, which Buchbinder and McCrone (2022) refer to as *teaching mathematics* via *reasoning and proving*. Their perspective is based on a set of principles for guiding mathematics teachers' classroom practice:

> (1) Reasoning and proving must be fully embedded in the existing mathematics curriculum; (2) Emphasis should be made on deductive reasoning in knowledge production and validation, while (3) Using language, notation, and representations within the conceptual reach of the students.
> *(p. 2).*

Stylianides and Ball (2008) discussed two interrelated forms of knowledge that can be used to cultivate proving in classrooms: knowledge of different kinds of proving tasks and the relationship between proving tasks and proving activities.

However, despite the acknowledgement of its importance, in many countries, it seems to have stayed "understated", also in national curricula, until relatively recently, when there has been an upsurge in those national documents and textbooks (e.g., in Norway, Slovakia, and the Netherlands). Unfortunately, the (re-)emergence has not always made its way into national classrooms and hence needs to be addressed in teacher professional learning activities. The *Principles and Standards for School Mathematics* (NCTM, 2000) emphasise the importance of students developing and being able to assess mathematical reasoning from pre-kindergarten through grade 12. For students in grades 6 to 8, R&P skills introduced in earlier years should be further refined, focusing on using inductive and deductive reasoning to construct mathematical arguments.

2.2.1 Characterisations and roles of reasoning and proving in education

Mathematical reasoning has several definitions or characterisations. Bergqvist and Lithner (2012, p. 253) define it in the following way:

> The line of thought that is adopted to […] reach conclusions when solving tasks. The reasoning is not necessarily based on formal logic and is therefore not restricted to proof; it may even be incorrect as long as there are some sensible (to the reasoner) reasons supporting it.

Battista (2016) understands mathematical reasoning as a "process of manipulating and analysing objects, representations, diagrams, symbols, or statements to draw conclusions based on evidence or assumptions" (p. 1). Similarly, Jeannotte and Kieran (2017) understand mathematical reasoning as a "process of communication with others or with oneself that allows for inferring mathematical utterances from other mathematical utterances" (p. 9). The NCTM (2009) characterises reasoning as an activity which "involves drawing logical conclusions based on evidence or stated assumptions". On the webpage of NRICH, provided by the University of Cambridge team, reasoning is said to enable children "to make use of all their other mathematical skills and so reasoning could be thought of as the 'glue' which helps mathematics makes sense" (The NRICH Primary Team, 2014).

According to Stylianides (2007), a proof in school mathematics is defined as a *mathematical argument*, a connected sequence of assertions for or against a mathematical claim, with the following characteristics:

1. It uses statements accepted by the classroom community (*set of accepted statements*) that are true and available without further justification;
2. It employs forms of reasoning (*modes of argumentation*) that are valid and known to, or within the conceptual reach of, the classroom community; and
3. It is communicated with forms of expression (*modes of argument representation*) that are appropriate and known to, or within the conceptual reach of, the classroom community. (p. 291).

While this definition focuses on the end product of mathematical exploration, Stylianides emphasises that other mathematical activities and forms of reasoning often precede and underpin proving. These include different modes of reasoning, such as inductive reasoning, deductive reasoning, reasoning by analogy, or experimental demonstration.[1] These modes play a crucial role in developing and justifying mathematical arguments that may eventually lead to valid proofs.

Stylianides (2008, p. 9) introduced the hyphenated term reasoning-and-proving to capture this integrated process, encompassing identifying patterns, making conjectures, providing non-proof arguments, and providing proofs. Similarly, in Jeannotte and Kieran's (2017) model of mathematical reasoning, the process aspect refers to processes related to the search for similarities and differences, as well as to processes related to validating, which includes justifying, proving, and formal proving. Cai and Cirillo (2014) highlight how researchers increasingly view R&P as a unified construct. At the same time, R&P are related: reasoning can occur without proving, but proving is difficult to carry out without reasoning (p. 139).

In terms of roles, de Villiers (2012) identified and described six key purposes of proof, which provide a broad framework for understanding the roles of R&P in mathematics. According to de Villiers, these purposes are: (i) **verification**, focused on confirming the truth of a statement; (ii) **explanation**, which offers insights into why a statement is true; (iii) **discovery**, which involves the identification or creation of new rules; (iv) **systematisation**, which organises various findings into a coherent deductive system of axioms, fundamental concepts, and theorems; (v) **communication**, which refers to the sharing of mathematical knowledge; and (vi) **intellectual challenge**, which captures the personal fulfilment and self-realisation that comes from constructing a proof. The six purposes of proof outlined by de Villiers (2012) shed light on the multifaceted role of R&P in mathematics education. These purposes (verification, explanation, discovery, systematisation, communication, and intellectual challenge) highlight how R&P intertwine cognitive, communicative, and epistemological processes. This multidimensionality reinforces R&P's position as a cornerstone of mathematical learning and teaching, fostering both understanding and application.

Integrating R&P into classroom practice requires careful attention to both task design and pedagogical strategies. Tasks that encourage identifying patterns, making conjectures, and constructing arguments help students engage meaningfully with R&P, promoting critical thinking and deep learning. Moreover, fostering an environment where students can share their reasoning, challenge ideas, and collaboratively refine proofs cultivates not only mathematical competencies but also a sense of intellectual curiosity and empowerment. As teachers embrace these roles of R&P, they equip students with the tools to approach mathematics as a dynamic, reasoned, and evidence-based discipline. Embedding R&P-related practices into teacher education programmes can foster critical thinking, improve lesson planning quality, and encourage reflective teaching practices.

2.2.2 PMTs' beliefs about reasoning and proving

Research reveals several challenges teachers and PMTs face in effectively teaching R&P in mathematics. Studies, such as Morris (2007), show that PMTs often struggle to distinguish between formal mathematical proofs and informal reasoning, such as inductive arguments or example-based reasoning. For example, many PMTs mistakenly accept examples as sufficient proof, a misconception that can distort their understanding of what constitutes valid mathematical reasoning. Similarly, Buchbinder and Zaslavsky (2011) found that participants included in a study (10th grade students and teachers) tend to rely on intuition, have difficulties justifying patterns, lack deep analytical reasoning, and struggle with transitioning from descriptive to formal proof-based reasoning.

Lesseig (2016) found that teachers generally value proof as a means for building deep understanding, yet they also struggle with what makes a formal proof, often confusing it with illustrative examples. This confusion highlights the need for targeted support in learning to differentiate between empirical and deductive reasoning. Ko (2010) underscores this point, showing that PMTs frequently find empirical arguments more convincing than formal deductive proofs, leading to a preference for example-based reasoning over logical deduction.

Harel and Sowder (2007) also identified that both PMTs and experienced teachers sometimes struggle to distinguish between valid and invalid arguments, suggesting a gap in foundational understanding that affects both novice and seasoned teachers. Lesseig et al. (2018) explored PMTs' purposes for teaching proof and noted challenges with the abstract nature of proof concepts, particularly in making them accessible and relevant to students.

Herbert et al. (2015) provided further insight into how primary teachers perceive mathematical reasoning as a multifaceted process. Their study highlights that primary teachers see reasoning as involving several overlapping dimensions:

1. **thinking** – teachers view reasoning as a process of deep thinking, where students actively engage with mathematical concepts rather than passively receiving information.
2. **communicating thinking** – reasoning is seen as a medium for students to articulate their mathematical ideas, which reinforces their understanding and allows teachers to assess their reasoning processes.

3 **problem-solving** – reasoning also aligns with problem-solving, emphasising the role of reasoning in tackling unfamiliar problems and devising strategies.
4 **validating thinking** – teachers believe that reasoning should help students assess the accuracy of their thoughts and ideas, encouraging critical self-reflection.
5 **forming conjectures** – reasoning encourages students to make and test hypotheses, fostering a spirit of inquiry and exploration in mathematics.
6 **using logical arguments to validate conjectures** – teachers identify logical reasoning as essential for establishing the validity of conjectures, underscoring the importance of structured, deductive thinking.
7 **connecting mathematical aspects** – reasoning is seen as a tool for linking different mathematical ideas, facilitating a more integrated understanding of the subject.

(ibid., p. 27)

These dimensions highlight that R&P extends beyond mere verification; it includes communication, exploration, and synthesis, positioning R&P as a multifaceted tool for learning. The findings of Herbert et al. (2015), along with studies like those by Ko (2010), Lesseig (2016), and Lesseig et al. (2018), suggest that teachers need support not only in understanding formal proofs but also in appreciating and leveraging the diverse roles R&P can serve in the classroom. Targeted professional development that emphasises both empirical and deductive reasoning could help teachers cultivate critical thinking and mathematical understanding in their students.

2.2.3 Teaching (and learning) of reasoning and proving

Yopp (2009) offers valuable insights into this area by presenting strategies for distinguishing and teaching inductive and deductive reasoning in middle school classrooms. By contrasting these reasoning types through examples, Yopp provides a foundation for PMTs to understand R&P's complexity and to consider how different reasoning types contribute to mathematical proof. Buchbinder and Zaslavsky (2011) underscore the importance of transitioning students from intuitive reasoning to a more rigorous, analytical approach, highlighting a critical aspect of R&P in mathematics education by introducing non-routine problems into teaching practice.

Frameworks for different types of R&P have been proposed by various scholars. Stacey and Vincent (2009) proposed a framework for analysing R&P tasks consisting of seven categories of reasoning. This framework was used by

Silverman and Even (2015), who analysed 7th-grade textbooks and identified predominant empirical reasoning tasks. Built on that, a framework (Sevinc et al., 2022) was developed for analysing solved example tasks in the 8th-grade textbooks in five MaTeK countries.

Buchbinder and McCrone (2020) proposed the "Mathematical Knowledge for Teaching Proof" (MKT-P) framework, which draws from socio-cultural theories of learning (e.g., Lave & Wenger, 1991. This framework emphasises the specialised knowledge teachers need to teach proof, divided into three core knowledge domains specific to teaching proof: (i) content knowledge of proof (logical aspects of proof), (ii) knowledge of content and students specific to proof (KCS-P), and (iii) knowledge of content and teaching specific to proof (KCT-P). This division highlights that teaching proof is not only about understanding formal logic but also requires sensitivity to how students think about and misunderstand proof and pedagogical strategies to bridge these gaps effectively. Through this lens, MKT-P positions proof instruction within the physical and social contexts of learning, highlighting the role of interaction and active participation in developing proof-related knowledge.

Similarly, Stylianides (2008) emphasises a three-part framework encompassing mathematical, psychological, and pedagogical aspects. The "Analytical Framework for R&P" offers a structured approach to understanding these dimensions, illustrating the interplay of knowledge and cognitive processes necessary for teaching R&P effectively. Later on, Stylianides and Stylianides (2022) offer a compelling application of theoretical frameworks for teaching proof through their work on a structured learning trajectory aimed at both students and PMTs. Their study, developed through two multi-year design experiments, proposes a trajectory with two key milestones: (1) *seeing a need to learn about proof*, and (2) *developing an operationally functional conceptualisation of proof*.

The first milestone, *seeing a need to learn about proof*, serves a dual purpose. For students, it involves epistemological justification – understanding why proof is essential for establishing mathematical truth. For PMTs, it also includes pedagogical justification, equipping them to convey the importance of proof in their future classrooms. This stage is centred on cultivating a sense of purpose and relevance for proof, forming a foundational motivation that encourages engagement with more complex proof concepts.

The second milestone, *developing an operationally functional conceptualisation of proof*, guides learners from simple logical statements to progressively complex proofs, reinforcing an understanding of logical sequencing and rigour. Through this stage, students and PMTs alike gain hands-on experience in constructing arguments, allowing them to move beyond rote learning towards a more robust, practical comprehension of proof.

Finally, Ellis (2011) focuses on teacher moves that support students' mathematical R&P skills, particularly in contexts where students engage in generalisation and justification. In later work, Ellis et al. (2019) provided a refined framework for supporting student reasoning through teacher moves, focusing specifically on inquiry-oriented instruction. This framework, known as Teacher Moves for Supporting Student Reasoning (TMSSR), organises pedagogical strategies into four categories: eliciting, responding, facilitating, and extending. Together, these studies provide complementary insights into effective teaching practices for enhancing R&P in mathematics classrooms.

The mentioned studies highlight the multi-dimensional nature of teaching R&P and the need for specific frameworks and targeted interventions. They underscore the importance of teacher knowledge in content, pedagogy, and didactics, as well as the role of classroom experiences in developing effective R&P teaching skills.

2.3 Research questions

In order to be able to pursue our overall research goal (how can PMTs develop lesson design capacity), we designed a specific intervention (denoted here with capital I) for teacher education and asked the following research questions:

RQ1: What are PMTs' beliefs concerning R&P in teaching before and after the Intervention, and how can we understand these, considering that PMTs worked in five different (national and cultural) contexts?

RQ2: What are the characteristics of and the similarities and differences among the different lesson plans? In which ways did lesson plans change/progress during the Intervention, and how was PMTs' progress influenced by the set-up of the Intervention?

RQ3: How did the PMTs experience the Intervention?

Note

1 Experimental demonstration engages students in some form of enactive participation, examining one or more physical or numerical instances (Stacey & Vincent, 2009, p. 280).

3

BACKGROUND OF THE STUDY

In this chapter, in the first part, we provide detailed accounts of the context of mathematics teacher education in the five "MaTeK countries", with a particular focus on how lesson planning and reasoning and proving (R&P) are implemented in teacher education in MaTeK institutions. In the second part, we focus on R&P in the intended curriculum in the five countries. In both sections, we only consider teacher education and curriculum related to the school grades in which the participants of the intervention acquired their teaching qualification, namely 6–13 in the Czech Republic and Italy, 5–10 in Norway, 5–13 in Slovakia, and 5–12 in Türkiye. This choice ensures that the description of teacher education and curriculum is directly aligned with the professional scope of the pre-service mathematics teachers (PMTs) involved in the study.

3.1 Context of teacher education

In general, in the mathematics education course in the Master's programme, PMTs' pedagogical content knowledge and subject matter knowledge (Ball et al., 2008) are developed, strongly linked to the current research, as well as the national mathematics curriculum.

3.1.1 Czech Republic

All teachers in the Czech Republic must have a Master's degree. They are educated in state universities only, at the faculties of education and other

DOI: 10.4324/9781003533870-3

faculties (e.g., the faculties of science, the Faculty of Mathematics and Physics). Teachers usually study two subjects (e.g., mathematics and physics, mathematics and physical education, mathematics and the German language, etc.). At Charles University (CU), there is also a study programme for mathematics teachers only (they get twice as many mathematics and mathematics education courses and teaching practice). Both types of teachers have the same qualifications as teachers of 11–19-year-old pupils and work in the same schools.

The proportion of individual parts of the study (i.e., mathematics, mathematics education, psychology and education, teaching practice) is given by the Ministry of Education, Youth and Sports. The faculties suggest the content of the courses, and the programme is accredited either by the National Accreditation Bureau for Higher Education or by universities. There is a great deal of tradition in the system of mathematics courses, and the programmes at different universities do not differ vastly.[1]

Previously, there had been a five-year undivided study for future teachers; nowadays, such a study only exists for future elementary teachers. Lower and upper secondary teachers study a three-year Bachelor's study followed by a two-year Master's study. The Master's study finishes with the final state examination at the faculty (there is no national examination) and the defence of a thesis. The final state examination is an oral examination in mathematics, mathematics education, psychology, and pedagogy. The Master's thesis usually takes the form of a small experimental study and is about 70–100 pages long.

In this study, we worked with a group of PMTs from CU who had completed a three-year Bachelor's degree and one year of the Master's study. They were beginning their second year of the Master's study. Award of the Bachelor's degree was based on students' successfully completing courses conferring a total of 180 academic credits; the Master's degree, a further 120 credits.[2]

In their Bachelor's degree, they had completed:

- 62 credits in mathematics courses
- 7 credits in mathematics education courses
- 2 credits in assistance practice in mathematics lessons
- 71 credits in courses related to their chosen second school subject
- 18 credits in general pedagogy and psychology
- 11 credits for writing the Bachelor's thesis
- 9 credits in vocational courses (they can choose any courses at the University)

During their Master's study, they must gain the following credits (the number in square brackets shows the number of credits they had completed before the intervention, which was organised during their second year of the Master's study):

- 16 credits in mathematics courses [7]
- 13 credits in mathematics education courses [10]
- 2 credits for observation practice in mathematics [2]
- 8 credits for teaching practice in mathematics [4]
- 39 credits in courses related to their chosen second school subject
- 4 credits for pedagogical and psychological practice with reflection [4]
- 19 credits in general pedagogy and psychology [8]
- 13 credits for writing the diploma thesis
- 6 credits for vocational courses

In terms of teaching practice, during their Bachelor's study, PMTs assist mathematics teachers in their tasks, such as assisting during the lessons, preparing tests and homework, marking tests, and evaluating homework. They complete 12 45-minute units of work at schools and 12 reflection lessons at the university.

In the first year of their Master's studies, PMTs observe approximately 30 mathematics lessons during observation practice and participate in 12 reflection lessons at the university. During the teaching practice at lower secondary schools, they spend a month teaching lessons (about 12 mathematics lessons and 12 lessons in the other subject they qualify for) and participate in 12 reflection lessons at the university. The same process is repeated during the teaching practice at upper secondary schools in the second year of their Master's studies. Moreover, in their Master's study, they spend one day a week at the primary or secondary school observing lessons (not necessarily in mathematics) and reflecting on the psychological and general educational matters with a teacher educator.

PMTs get some experience with lesson planning in the course on general education and school pedagogy (designing a lesson plan with an emphasis on educational goals and their alignment with pupil activities and the assessment of learning outcomes is one of the seven goals of this 4 ECTS credit course) and mathematics education courses (the emphasis is on planning the mathematics content for lessons). PMTs should have ample experience with mathematical proofs from mathematics courses. In terms of teaching R&P, there had been no explicit focus on teaching R&P in their previously completed mathematics education courses. Some proofs and reasoning behind, for example, mathematics formulas were covered together when the teaching of the content in question was addressed in mathematics education courses (e.g., the proof of the Pythagorean theorem when the ways of teaching this theorem were covered, or how to derive the formulas for measure in geometry when the course targeted measurement in geometry).

3.1.2 Italy

In Italy, all teachers for grades 6–13 must have a Master's degree and a specialised course degree for teaching. The Bachelor's and Master's programmes for grades 6–13 follow national regulations that specify minimum content requirements. However, these programmes, which are 3+2 years in length, may differ slightly between institutions, though not significantly.

For their Master's, PMTs can attend specialised courses (e.g., in mathematics or physics). Several universities also have a study programme for mathematics teachers that the university itself defines, and so the number of credits or types of courses offered to PMTs could differ.

Mathematics teachers for grades 6–8 are not specialised in a single subject; their university studies also focus on science teaching. As a result, lower secondary school teachers commonly teach both mathematics and science to 11–13-year-old students.

Teachers for grades 9–13 typically specialise in one subject (e.g., mathematics), with some courses related to other subjects (such as physics, computer science, etc.). The University of Palermo (UNIPA) also has a study programme specifically for mathematics teachers at this level.

In this study, we are particularly referring to this kind of teacher preparation course. We worked with a group of PMTs from UNIPA for grades 6–13 who had completed a three-year Bachelor's degree in mathematics. They were in their first or second year (beginning phase) of a two-year Master's degree programme.

In their Bachelor's degree, they had completed 180 credits distributed as follows:

- 108 credits in mathematics courses
- 6 credits in the history of mathematics course
- 18 credits in courses related to physics
- 9 credits in computer science
- 24 credits in courses focused, for example, on financial mathematics, computer science, statistics, pedagogy, and psychology
- 6 credits in the second language course (English, B1)
- 6 credits for skills useful for entering working life (PMTs can choose training at school)

After the first three years, PMTs have to pass a formal exam (3 credits).

During their Master's study, PMTs must gain 120 credits as follows:

- 48 credits in mathematics courses
- 12 credits in the history and mathematics education course (6 ECTS are optional)

- 6 credits in courses related to physics
- 30 credits in courses on, for example, pure mathematics, financial mathematics, computer science, statistics, mathematics education, probability, cryptography, pedagogy, and psychology
- 3 credits in the second language course (English, B2)
- 3 credits for skills useful for entering working life (PMTs can choose training at school)

PMTs must pass a formal exam (24 credits) at the end of their study programme.

In terms of R&P, PMTs have ample experience with mathematical proofs based on several mathematics courses, but these are often not explicitly focused on teaching R&P. During the Master's course on mathematics education, PMTs explore the teaching aspects of conjecturing, arguing, justifications, and formal proofs related to various school mathematics topics.

With respect to lesson planning, PMTs have only limited experience. Their basic competencies in lesson planning are addressed in the Master's course on history and mathematics education, which is scheduled after their practical training periods. Consequently, they have little opportunity to apply lesson planning knowledge during practice.[3]

Regarding practice, in the third year of their Bachelor's study, PMTs have the option to attend lower and upper secondary school lessons. Within this training, PMTs assist mathematics teachers during lessons, help prepare tests and homework, evaluate students, and so on. This training falls under the elective category "skills useful for entering working life" in the programme description, where students can choose school-based training or other forms of practical experience. For this reason, the training at schools is not compulsory. In their Master's study, PMTs can likewise choose practical training in different contexts, such as industry or other companies. If they opt for school-based training, they typically follow the class tutor's instructions or observe lessons rather than engaging in lesson design.

After completing their Master's degree and final state examination, prospective secondary school teachers in Italy must attend specialised training programmes. Previously regulated by *Scuola di Specializzazione Interateneo per la Formazione degli Insegnanti di Scuola Secondaria* (Inter Academic Graduate School of Secondary School Teacher Training) until 2009, teacher training is now overseen by the *Tirocinio Formativo Attivo (Active Training Internship)* programme, which includes coursework in pedagogy, psychology, and didactics, as well as 400 hours of supervised school training.

Reforms (MIUR, 2025) require future teachers to acquire 60 additional credits in qualifying courses called *PF60*. These include several courses in pedagogy, psychology, and assessment theories, for example. Such general education courses focus on teaching disciplines and school practice reflection (e.g., in mathematics education), 180 hours of mandatory school training, and a final qualifying examination.

3.1.3 Norway

In Norway in 2009, the former four-year general teacher education programme for grades 1–10 was divided into two separate programmes for grades 1–7 and 5–10. Since 2017, all primary (grades 1–7) and lower secondary (grades 8–10) school teachers have undertaken a five-year integrated Master's programme, "Master i grunnskolelærerutdanning", MGLU 1–7 or MGLU 5–10. These programmes follow national regulations that specify the minimum programme content requirements. While numerous requirements are in place, both institutions and PMTs have considerable flexibility in determining which subjects to provide and select. However, the programmes in different institutions do not differ significantly. When enrolling in an MGLU 5–10 programme, students must choose between mathematics or Norwegian as their core subject for the first phase of their studies. Later, for the Master's degree, student teachers can choose to do their Master's in a school subject (like Norwegian, mathematics, music, natural science, etc.) or in social or special pedagogy.

In this study, we worked with a group of student teachers in the MGLU 5–10 programme at the Norwegian University of Science and Technology (NTNU), specialising in mathematics. That means that their core subject, both in their first three years and in their Master's, was mathematics. These PMTs were in their fourth year of study. The programme overview of the first three years is as follows:

- 5 days of observational practice in the first year
- 80 school-based days of practice (25 in the first year, 30 in the second, and 25 in the third year)
- 30 credits in general pedagogy
- 60 credits in mathematics education courses
- at least 30 credits in courses related to PMTs' chosen second school subject (A)
- at least 30 credits in courses related to their chosen third school subject (B) or in special/social pedagogy
- an additional 30 credits in further studies in school subjects A or B, or in special/social pedagogy

In the second half of the third year, PMTs write a "Research and Development" thesis, integrating pedagogy and didactics of a subject A or B with experiences from practice. This thesis must be approved before students can formally work on their Master's thesis. As the programme is integrated, there are no state examinations after the first three years.

In the last two years of the Master's programme, PMTs must complete

- 30 school-based days of practice (15[4] + 15 days)
- 30 credits in general pedagogy
- 90 credits in mathematics education courses (including a Master's thesis of 30 credits)

There is no final state examination at the end of the study; PMTs (only) submit a Master's thesis, which two evaluators evaluate. After this, they are qualified teachers.

In terms of teaching practice, although our study's PMTs specialised in mathematics, they did not necessarily teach mathematics in their school placement. It depended on the specialisations of their assigned in-service teacher and their school-based mentor during teaching practice. While efforts were made to match students and the mentor regarding sharing the same subjects, this was not always possible.[5] PMTs conducted practice in mixed groups of 2–4 students of various specialisations (e.g., Norwegian, music, science). They planned and taught lessons together under their mentor's supervision, and in some cases, they taught alone.

In general terms, the main themes of practice in particular study years are, for example, building teacher identity, multicultural classroom, inclusive education, pupils with special needs, varied and adapted teaching, assessment for learning, school as an organisation, cooperation with parents, and other central bodies outside the school. As the focus is not on the school subjects, in the student teachers' education, lesson planning is mainly addressed in the pedagogy courses, utilising the "Didactical Relationship Model" (Hiim & Hippe, 2006) that is widespread in Northern European didactics. Although the model consists of six elements (learning conditions, setting, goals, content, learning process, assessment), teacher education institutions often provide the student teachers with a simplified template for lesson planning, addressing three questions: What? (content), How? (learning activities and teaching methods), Why? (justification of choices). The prepared lesson plans are discussed with the mentor, occasionally with teacher educators (of various specialisations) who are assigned to follow a group of students in practice school for one day per practice placement.

As listed earlier, during the first phase of their study, PMTs only took mathematics education courses, where the emphasis was given to the development of PMTs' pedagogical content knowledge for teaching school mathematics in grades 5–10. While mathematics lesson planning was not a central focus of these courses – appearing, at most, occasionally in assignments – the main orientation was towards aligning their teaching with current research results and the current mathematics curriculum and utilising exploratory, experimental, problem-solving methods, and investigative activities. In terms of R&P, in the courses, PMTs worked with generalisations, conjecturing, argumentation, justifications, and proving related to school mathematics topics (like figural patterns, odd and even numbers, prime numbers, or area, perimeter, and volume of geometric figures or objects). The proving methods utilised included mainly visual representation proofs and generic examples. PMTs experienced formal proofs and formal logic to a limited extent within a course focusing on school geometry, where they learned about Euclidean geometry and the Euclidean axiomatic system.

At the Master's level, PMTs developed knowledge of various theoretical perspectives and methods in research in mathematics education, also needed for writing their Master's theses. The intervention was conducted within the "Learning and teaching of mathematics" course, where learning theories were at the core, applied in the context of school algebra and algebraic reasoning. As the programme structure shows earlier, PMTs did not take stand-alone university mathematics courses; their study of mathematics was embedded within mathematics education courses throughout their programme.

3.1.4 Slovakia

All Slovak teachers must have a Master's degree. In most cases, they follow a study path consisting of a three-year Bachelor's programme and a two-year Master's programme. Teachers who teach pupils in grades 5–13 typically study two subjects (e.g., mathematics and physics; mathematics and biology). The study programmes offered for mathematics teachers do not differ significantly between Slovak institutions. At the Bachelor's level, the focus is primarily on mathematical content, while at the Master's level, the emphasis shifts to the didactics of mathematics. The study plan is usually based on a modular (block) system to allow PMTs to continuously adjust their focus within the chosen field of study and expand their academic scope. A study plan is divided into compulsory, compulsory elective, and elective courses. Both Bachelor's and Master's studies conclude similarly: PMTs must write a thesis, defend it, and pass state examinations.

In this study, we worked with a group of PMTs from Comenius University in Bratislava (UK BA), who had completed a three-year Bachelor's degree. The intervention took place in the first semester of their Master's programme. During their Bachelor's studies, they completed compulsory courses, compulsory elective courses, and elective courses. Because they could choose from a list of courses for the compulsory elective and elective courses, we provide next the minimum number of credits that each PMT had to complete:

- 42 credits in compulsory courses in mathematics
- min. 8 credits in compulsory elective courses in mathematics
- 1 credit for observational practice in mathematics
- 2 credits for the state exam in mathematics
- about 46 credits in compulsory courses, including practice, related to their chosen second school subject
- about 2 credits in compulsory elective courses related to their chosen second school subject
- 2 credits for the state exam in the second subject
- 25 credits in compulsory courses in pedagogy and psychology
- min. 12 credits in compulsory elective courses in pedagogy and psychology
- 8 credits for Bachelor's thesis

Within the Master's programme, the distribution of the credits is as follows:

- 20 credits in mathematics education compulsory courses
- min. 6 credits in compulsory elective mathematics education courses
- min. 6 credits in compulsory elective mathematics courses
- 5 credits for teaching practice in mathematics
- 3 credits for the state exams in the didactics of mathematics
- 20 credits in compulsory courses related to their chosen second school subject
- min. 6 credits in compulsory elective second subject courses
- 5 credits for teaching practice in the second subject
- 3 credits for the state exams in the didactics of second subject
- min. 13 credits for compulsory courses in general pedagogy and psychology
- 10 credits for Master's thesis

In terms of teaching practice, PMTs had completed it at both lower and upper secondary schools. During their Bachelor's studies, PMTs undertook a week-long observational practice in their third year. During this week, they worked with the same in-service teacher, observing their teaching methods, techniques, and how they handled various situations inside and outside the mathematics classroom. PMTs had to observe 15 mathematics lessons during this practice and were not expected to teach any lessons. In the Master's programme, PMTs participated in a two-week practice in the first year and a three-week practice in the second year of their studies. During each teaching practice, PMTs worked under the supervision of an assigned in-service teacher. Within this context, they participated in teaching, designing homework and tests, and grading students' work. Lesson planning had been introduced earlier in general pedagogy courses (Bachelor's level), but only at a general and theoretical level. PMTs' first opportunity to plan lessons in a school-based setting occurred after the intervention (Master's level). During this practice, they could consult their lesson plans with the assigned teacher before teaching and receive feedback on their implementation afterwards. In the first year, PMTs were required to participate in 20 lessons, teaching at least ten of them under the teacher's supervision. In the second year, they had to participate in 30 lessons, teaching at least 15 of them under supervision.

Regarding R&P, during the Bachelor's study, PMTs' subject matter knowledge related to advanced mathematics in algebra, calculus, geometry, probability, and statistics was developed, with a major component being statements and their proofs. Therefore, PMTs encountered various types of reasoning and proof (e.g., direct, indirect, proof by contradiction) but had not yet learned how to support R&P in mathematics teaching. The aim was to teach them not only to understand proofs but also to be able to create them independently. In the Master's study, the emphasis was on developing PMTs' pedagogical content knowledge for teaching school mathematics.

3.1.5 Türkiye

The lower and upper secondary school mathematics teacher education programmes in Türkiye are four-year undergraduate programmes. The programmes are both under the Faculty of Education, Mathematics and Science Education Department. In 2018, the Council of Higher Education (CoHE, 2018) updated all teacher education undergraduate programmes. These programmes were divided into three categories: professional teaching knowledge, pedagogical content knowledge, and general culture courses (Avşar Tuncay, 2022). Following this update, most universities updated their teacher education programmes, including Middle East Technical University (METU), to reflect the changes in courses and credit allocations. In 2020, CoHE empowered the appropriate boards at higher education institutions to make decisions regarding the curriculum in their academic programmes (Avşar Tuncay, 2022). Following this decision, the undergraduate lower and upper secondary school mathematics teacher education programmes at METU were updated again for Fall 2020.

The undergraduate programmes (both lower and upper secondary school mathematics teacher education programmes) at METU include courses such as mathematics and physics (courses are offered by METU's Faculty of Arts and Science), general pedagogy courses such as "Introduction to Education" (provided by METU's Educational Sciences Department at the Faculty of Education), and pedagogical content knowledge courses such as "Introduction to Mathematics Education" (provided by METU's Mathematics and Science Education Department at the Faculty of Education).

For the lower secondary school mathematics teacher education programme, the PMTs completed about 267 credits in total. The credit allocations are as follows:

- 77 credits from the Faculty of Arts and Science, 55 of which are mathematics courses
- 25 credits from the Educational Sciences Department at the Faculty of Education
- 57 credits from the Mathematics and Science Education Department at the Faculty of Education
- Practice Teaching I and II: 10 credits each in the last two semesters from the Mathematics and Science Education Department at the Faculty of Education

For the upper secondary school mathematics teacher education programme, the PMTs complete about 274 credits in total. The credit allocations are as follows:

- 110 credits from the Faculty of Arts and Science, 86 of which are mathematics courses
- 15 credits from the Educational Sciences Department at the Faculty of Education
- 56 credits from the Mathematics and Science Education Department at the Faculty of Education
- Practice Teaching I and II: 10 credits each in the last two semesters from the Mathematics and Science Education Department at the Faculty of Education

In both programmes, PMTs also take courses such as academic English, Turkish, computer applications or programming, and career planning. They also need to take several elective courses.

The medium of instruction at METU is English, so students take an English Preparatory Class when enrolling in an undergraduate programme at METU.

The Practice Teaching courses are generally offered in the last year for two semesters. In that course, PMTs attended school for six lesson hours each week, for 12 weeks each semester. They were assigned a lower or upper secondary school mathematics teacher at the school (Practice Teacher) and had duties such as preparing and implementing mathematics activities and lesson plans, developing assessments, and evaluating student work. The practice teaching course instructor and Practice Teacher observed each PMT during two mathematics lessons (one at the beginning and the other towards the end of the semester) and evaluated the PMT's lesson planning and implementation based on the criteria provided by the Ministry of National Education. PMTs learn about lesson planning in the first course they take from their own department in their fourth semester. Later, they prepare and submit lesson plans in several courses, including mathematics teaching method courses and practice teaching courses.

The three-year graduate programme, Mathematics Education, under METU's Faculty of Education, includes seven courses: three are mandatory (Educational Statistics, Research Methods and Ethics, and Fundamentals of Research in Mathematics Education I, which are 8 ECTS each), and four are elective. Candidates should also complete a seminar course of 10 ECTS. After completing their coursework within a maximum of two years, candidates should successfully defend their Master's thesis.

To be certified as a lower and upper secondary school mathematics teacher, an undergraduate diploma is required in Türkiye. After graduation, to be appointed as a mathematics teacher in public schools, teacher candidates must pass the National Public Personnel Selection Examination and be interviewed by officials from the Ministry of National Education.

The intervention for this study took place within the "Projects in Mathematics and Science Education" elective course. This elective course is offered to fourth-year and Master's-level students, in which students develop and implement an independent research project with the instructor's guidance and submit a project report at the end of the semester. However, the plan for the intervention was mentioned to the students in the first week of the course, and the student-initiated projects started only after the intervention.

The PMTs learn about formal proofs in various mathematics courses (e.g., Fundamentals of Mathematics, Elementary Geometry), which they mostly complete in their first three years from the Mathematics Department under the Faculty of Arts and Science. Typically, PMTs have not specifically focused on teaching and learning R&P in their previous courses.

3.1.6 Summary

The curriculum for mathematics teacher education differs across the five MaTeK institutions. Table 3.1 provides an overview of the similarities and differences among the MaTeK institutions, with a focus on courses and the number of credits, which PMTs participating in the study took **prior to the intervention**.

While the programmes share core elements in preparing PMTs with relevant subject knowledge, pedagogical skills, and practical experience, the approach and emphases vary significantly, reflecting each country's educational priorities and structure. For example, Norwegian PMTs completed 60 credits in mathematics education courses before the intervention, whereas Slovak and Italian PMTs had none. Conversely, all PMTs except the Norwegian

TABLE 3.1 Overview of PMTs' preparation in MaTeK institutions

	Czech Republic/CU	*Italy/UNIPA*	*Norway/ NTNU*	*Slovakia/ UK BA*	*Türkiye/ METU*
PMTs qualified to teach in grades	6–13	6–13	5–10	5–13	5–8 or 9–12
Mathematics education courses	17 credits	0 credits	60 credits	0 credits	57/56[a] credits
Mathematics courses	69 credits	114 credits	0 credits	52 credits	55/86[a] credits
Pedagogy/ Psychology courses	26 credits	24 credits	30 credits	37 credits	25/15[a] credits
Observational/ assistance practice	30 lessons/ 12 lessons	10 days/-	5 days/-	5 days/-	–
Teaching practice	12 lessons	10 days	95 days	0 days	0 lessons/72 lessons[b]

[a] The first number indicates the number of credits for PMTs studying to teach at the lower secondary school level, and the second is related to the upper secondary school level PMTs.

[b] The first number indicates the number of mathematics lessons for undergraduate-level intervention participants, and the second is related to the Master's level intervention participants.

group gained significantly more credits in mathematics courses, with Italian PMTs gaining the most: they completed double the number of credits than the Slovak or lower secondary school Turkish PMTs. This indicates a stronger focus on mathematical content knowledge.

Each institution, except for Türkiye, offered some form of observational or assistance practice before full teaching practice, typically with increasing responsibilities as PMTs progressed. NTNU emphasises extensive in-school practice, a total of 95 days before the intervention. The table also reflects zero totals for Slovakia and Türkiye (for undergraduate PMTs), as Slovakia's practical training was scheduled after the intervention. In Türkiye, it commenced during the same semester as the intervention. In Italy, teaching practice was optional for PMTs, suggesting that students from these three countries may have had less experience in lesson planning.

All educational systems are described in more detail in Eurydice (https://eurydice.eacea.ec.europa.eu/eurypedia)

3.2 Reasoning and proving in the national curriculum

3.2.1 Czech Republic

The Czech Republic has a national curriculum: the Framework Educational Programme for Basic Education (grades 1–9) and the Framework Educational Programme for Secondary Education (grades 10–13). Both documents briefly characterise education, mention educational conception and objectives, and include the mandatory number of lessons for different subjects. There is also a section dedicated to so-called key competencies describing what a pupil leaving the school should know, be able to do, or value.

The section called Mathematics and its application describes general learning objectives and mentions different educational content areas in mathematics (such as logic and proof, number and variable). The content areas are described using only so-called expected outcomes, stating briefly what a pupil should learn in a given area. Both documents are rather short, not providing teachers with any tips, suggestions, or ways to achieve the given goals. For instance, the part of the curriculum dedicated to mathematics consists of only eight pages for primary and four pages for secondary grammar school. Schools are presumed to prepare a detailed document called the School Educational Programme, which lists specific learning outcomes for each subject, that is, elaborates on what the Framework prescribes in far more detail.

Considering the R&P area, the Framework Educational Programme for Basic Education (MŠMT, 2023) includes three objectives related to it:

- developing combinatory and logical thinking, critical judgement, and comprehensible and factual argumentation by solving mathematical problems
- developing abstract and precise thinking by acquiring and using basic mathematical concepts and relationships, recognising their characteristic qualities, and identifying and classifying concepts based on these qualities
- learning to trust in one's own problem-solving skills and abilities, systematic self-evaluation at each step of the solution process, developing a systematic approach, determination, precision, and the ability to express hypotheses based on experience or experiment and to verify or reject them using counterexamples

The part dedicated to expected outcomes mentions R&P only when discussing reasoning in geometry: "[pupils will] reason and apply the positional and metric properties of basic two-dimensional figures when solving tasks and simple practical problems; use the necessary mathematical notation" and "apply theorems on congruent and similar triangles when making argumentations and calculations".

The Framework Educational Programme for Secondary Grammar Schools (MŠMT, 2023) includes the following objectives related to R&P:

- developing logical thinking and judgement, forming hypotheses based on experience or experiment, verifying or refuting them based on contradiction
- justifying mathematical strategies, defending one's strategy

One of the educational contents presented in the Framework is Logic and Proof, in which the pupil shall:

- read and record theses in the symbolic language of mathematics
- properly use logical connectives and quantifiers
- distinguish between definitions and theorems, discern premises and conclusions of the theorems
- discern between correct and incorrect judgements
- create hypotheses, justify their validity or invalidity, and refute invalid statements
- justify their approach and verify the validity of their steps in the solution of the problem

3.2.2 *Italy*

In 2012 (with an update in 2018), the Italian Ministry of Education, University and Research (MIUR) published the *National curricular guidelines for pre-primary school and for the first cycle of education*. This document, until 2025, represented a framework for schools in the design and implementation of local primary (grades 1–5) and lower secondary (grade 6–8) curricula. Schools are free to determine content and methods of instruction autonomously, provided they are consistent with the learning objectives established by the *National curricular guidelines*. The objectives and expectations of the mathematics curriculum at the lower secondary level are summarised according to four subject areas: Numbers, Space and Shapes, Relations and Functions, Measurement, Data, and Chance. In 2025, the Italian Ministry of Education published a draft version of the new guidelines for kindergarten and the first cycle of education.

The theme of R&P, expressed in the Italian mathematics curriculum for grades 1–13 (MIUR, 2010, 2012, 2025), is generally presented as a crucial theme that lends itself to considerations on interdisciplinary aspects and concerns, in particular the relationship between language and mathematics learning (Mariotti, 2006; Mariotti et al., 2018).

In detail, the Italian national curriculum for elementary and lower secondary education (grades 1–8) refers to R&P as a starting point for mathematical proving and as a process of argumentation, logical thinking, hypothesis testing, and critical evaluation in problem-solving. As regards upper secondary school education (grades 9–13), arguing, conjecturing, and proving is one of the three transversals, called "process", reported in the mathematics curriculum proposed by the Italian Mathematical Union (UMI) for secondary school (Anichini et al., 2004) in relation to four thematic nuclei: Arithmetic and Algebra, Relations and Functions, Space and Shapes, and Data and Predictions. These are also discussed in the 2010 National guidelines concerning the specific learning objectives for the activities and teachings included in the different curricula for grades 9–13, called *Indicazioni per i Licei* and *Linee Guida per gli Istituti Tecnici e Professionali* (MIUR, 2010), and in the framework of national standard tests INVALSI in mathematics (INVALSI, 2018).

UMI and the Italian Commission for the Teaching of Mathematics (UMI, 2013), to offer a useful tool to students and teachers, as well as to test writers, have drawn up a proposal for a syllabus of relevant mathematics topics for the *Liceo Scientifico* State Examination, in the light of the National Indications. An aspect of interest in these documents is "the fact that the explicit reference to demonstration as an object is only in the field of Geometry and with particular reference to Euclid's *Elements*" (Baccaglini-Frank et al., 2018, p. 162). While references to the proving aspects would, in fact, seem to be of a completely general nature (Durand-Guerrier et al., 2012).

3.2.3 Norway

The new Norwegian mathematics curriculum (LK20) for grades 1–10, which was launched in August 2020, is competency-based, defining 10 to 13 competence goals for each grade (starting from year 2). Every competence goal comprises a verb denoting competence and a noun clause outlining a specific mathematical domain. For instance, "explore and argue for formulas for area and volume of three-dimensional figures" (Ministry of Education and Research, 2019, p. 14). The verbs vary from *calculate, carry out, represent and use to explore, formulate, generalise, explain*, and *argue*.

In addition to the measurable goals for each grade, LK20 also introduced six process-oriented core elements, emphasising how pupils learn and engage with mathematical concepts. One is *Reasoning and argumentation*, which emphasises the importance of pupils understanding that mathematical rules and results have clear reasons. It encourages them to articulate their reasoning and justify their solutions to demonstrate validity.

In LK20, processes related to mathematical reasoning (Jeannotte and Kieran, 2017) are not included only under the core element *Reasoning and argumentation*, but also under the core elements:

- *Exploration and problem solving*, as "pupils search for patterns, find relationships and discuss their way to a shared understanding",

- *Representation and communication*, as "pupils shall have the opportunity to explain and give reasons for the choice of form of representation", and
- *Abstraction and generalisation*, as "pupils can explore numbers, calculations and figures to find relationships, and then formalise by using algebra and suitable representations"

(Ministry of Education and Research, 2019, pp. 2–3).

This indicates quite a strong emphasis on reasoning. However, the verb "prove" is mentioned in the LK20 curriculum only once, in the paragraph related to the core element *Reasoning and argumentation*: "Argumentation in mathematics means that the pupils give reasons for their approaches, reasonings and solutions, and prove that these are valid" (Ministry of Education and Research, 2019, p. 3). The notion of proof is not explicitly mentioned among the competence goals. Valenta and Enge (2020) analysed LK20 with regard to delineating competencies associated with proof. They concluded that there is very little focus on developing logical thinking, deductive reasoning, and working with definitions. Thus, the development of pupils' competence with proof and proof-related work is left to teachers and textbook authors. However, since the competence goals are rather wide and implicit, teachers' interpretation of R&P can vary and be challenging.

3.2.4 Slovakia

The national educational programme in Slovakia is divided into individual levels. This means that there is a national curriculum for lower secondary education (grades 5–9) and one for upper secondary education (grades 10–13). In the national educational programme, subjects are divided into educational areas. Mathematics belongs to the educational area called Mathematics and work with information. In lower secondary education, this area primarily focuses on active activities, such as working with objects and applying knowledge in real situations. This area, among other things, should develop pupils' logical and critical thinking and teach them to formulate hypotheses and support statements with arguments. This should lead to the precise expression of thoughts and procedures and their recording in formal notes, which should serve as a general means of communication (SPI, 2015a).

Looking closer at the characteristics of mathematics, one can see that it is primarily focused on building basic mathematical literacy and developing cognitive areas, including reasoning and solving more complex problems that require a broader understanding of connections and relationships (SPI, 2014a). Among the goals of the subject is that pupils are expected to develop their logical and critical thinking, reasoning, communication, and collaboration in a group to solve a problem.

Regarding upper secondary education (SPI, 2015b), the goal of the educational area of Mathematics and work with information is to develop mathematical thinking and deepen the pupil's abstract, analytical, systematic thinking, and logical reasoning. It should also teach pupils to argue comprehensively and substantively. Among the goals of the subject of mathematics (SPI, 2014b), as in lower secondary education, is for pupils to argue when solving a problem. It is emphasised that teaching should be guided to develop pupils' logical and critical thinking and their ability to argue. In contrast to lower secondary education, the educational content of mathematics is divided into five thematic areas, and one of these areas is Logic, Reasoning, Proof. This area includes topics such as statements, logical connectives, and proofs. Within the topic of statements and logical connectives, among the performance standards is the ability to "identify errors in arguments and reasoning".

3.2.5 Türkiye

In Türkiye, a national curriculum is implemented for school education. The lower secondary school mathematics curriculum (grades 5–8) consists of five learning areas: numbers and operations, algebra, geometry and measurement, data, and probability. The upper secondary school mathematics curriculum (grades 9–12) includes three learning areas, which are numbers and algebra, geometry, and data, counting, and probability.

Reasoning and proof are mentioned only in general terms in the lower secondary school mathematics curriculum. It states, "[The students] will be able to express their thoughts and reasoning easily in the problem-solving process and see deficiencies or gaps in others' mathematical reasoning" (Ministry of National Education [MoNE], 2018a, p. 9) under the specific aims of the mathematics curriculum for grades 5–8. In the high school mathematics curriculum (grades 9–12), terms such as propositions, definitions, axioms, proofs, theorems, and hypotheses are given under the Grade 9 topic "Logic" (MoNE, 2018b, p. 18).

In 2024, a new national mathematics curriculum was introduced for all levels. Reasoning and proof are given more emphasis in these curricula. For instance, mathematical reasoning is introduced as one of the content skills. It is defined to include analysing, interpreting, inferring, and mathematical verification and/or proof, as well as their process components. MoNE (2024a, 2024b) associate the themes explicitly with reasoning and proof when relevant.

3.2.6 Summary

The treatment of R&P across the national curricula of the five "MaTeK countries" reveals both shared priorities and notable differences. All five countries emphasise reasoning, logical thinking, and argumentation as important components of mathematics education, aiming to develop students' abilities to analyse, justify, and critically evaluate mathematical statements. These goals are generally interwoven with broader competencies, such as problem-solving and communication, rather than being isolated skills. Additionally, R&P is often associated with key mathematical domains like geometry and algebra, though to varying extents.

The depth and explicitness of R&P integration vary significantly. Norway's curriculum, for example, integrates reasoning through broad process-oriented goals but mentions "proof" only minimally. Italy's curriculum highlights R&P in several aspects, especially related to logic and geometry, yet leaves schools with considerable flexibility in implementation. On the other hand, Slovakia and the Czech Republic define R&P with more specificity, with Slovakia including a dedicated section on logic, reasoning, and proof at the upper secondary level, and the Czech curriculum addressing these through expected outcomes in logic and geometry. Türkiye shows a shift over time: while the older curriculum included limited references to R&P, the new 2024 curriculum places greater emphasis on mathematical reasoning, defining it as a skill that spans across content areas.

Notes

1 At present, the Competence Framework for Graduate Teachers has been prepared and will be used as a guide for teacher educators to follow.
2 Credits are all European Credit Transfer and Accumulation System (ECTS) credits. ECTS is the standard framework used across Europe to measure and compare students' workload in higher education. One ECTS credit corresponds to about 25–30 hours of student work (including lectures, assignments, projects, and independent study), and a full-time academic year equals 60 ECTS credits.
3 At the time of writing, steps are being taken to schedule the course before practice and involve PMTs in lesson planning during school placements.
4 The PMTs in our study had completed this practice just before the start of the Intervention.
5 A new model was meanwhile prepared, launching in the academic year 2024/25 to avoid this inconsistency.

4

METHODOLOGY

In this book, we use the term *intervention* in the sense of Stylianides and Stylianides (2013, 2017), who use it for any action taken to improve a situation in relation to the teaching and learning of mathematics at all levels of education, in our case, teacher education. Our intervention fulfils the basic requirements they place on education research that has good potential to support improvements in the classroom, including teacher education (Stylianides & Stylianides, 2017). First, the intervention is conducted in teacher education courses that are part of the university curriculum and in close cooperation between researchers and mathematics education course leaders, "thus increasing the likelihood that the results of the research will be directly applicable (instead of merely potentially relevant) to practice" (p. 334). Second, our intervention directly addresses problems of pre-service teachers' learning (namely, learning to design lessons) and how this learning can be better supported by teaching. Third, our intervention "seeks to develop empirically tested and theory-based solutions" and "also to explain how/why things 'worked'" (p. 334).

We collaboratively prepared the design of the intervention. We then applied it in teacher education courses at the five institutions, continuously discussing the design among ourselves and adjusting parts of the intervention if needed. Thus, our research uses a multiple case study design approach (Yin, 2009). Each case study consists of the implementation of the agreed intervention, applied in different contexts and with different populations. The advantage of using a multiple case study is that it allows for data analysis (a) within each context and (b) across different contexts. A multiple case study allows better generalisation than a single case study. For example, by considering our intervention in different cultural settings, we could distinguish

DOI: 10.4324/9781003533870-4

aspects that worked across such settings from those that would need adjustment.

Concerning the different sections of this chapter, we first explain the rationale for the different design features of the Intervention, including what these entailed. Second, we elaborate on the Intervention, describing and explaining the different sessions, the participants in each country, and the implementation differences in each context. Third, we explain the data collection strategies used and the data analyses for each research question.

4.1 Rationale for the design features of the Intervention

As stated in Chapter 2, there is a pressing need to enhance teachers' lesson design capacity to develop their pupils' reasoning and proving (R&P) skills in mathematics teaching. Thus, our Intervention aimed to help future mathematics teachers' lesson design related to the development of pupils' R&P.

The research team conducted an anticipatory thought experiment (Gravemeijer & Prediger, 2019), envisioning how the stated aim might be reached. This involved suggesting how it might be reached in one's own country and considering cultural differences between the countries. The team strove to find a common suggestion for the Intervention to be implemented in the five countries. This process was demanding and required multiple redesigns of the original suggestion before all the teacher educators felt they could implement the Intervention in their courses to benefit their PMTs.

Following the literature on teacher learning and considering that presenting PMTs with new knowledge related to teaching a subject matter through a lecture is ineffective, we decided to ground the Intervention in three design features. The first was reading professional texts, which were mainly used to convey the meaning of concepts related to R&P and the teaching process. The other two were based on the general acknowledgement that teachers' learning is most effective when grounded in experience. Situated cognition learning theory posits that learning should be rooted in authentic activity; that learning occurs within a community of individuals engaged in inquiry and practice; that more knowledgeable "masters" guide or scaffold the learning of novices; and that expertise is often distributed across individuals (Whitcomb, 2003, p. 538). Hence, in our Intervention, the practice-based design features consisted of working with mathematical tasks, planning and enacting a mathematics lesson, and reflecting on the plans and enactment.

4.1.1 Reading and reflecting on professional texts

Professional texts can be an effective means of getting acquainted with theoretical concepts that PMTs need to plan the teaching of a specific content effectively, either on their own or embedded in tasks for PMTs (such as the task of planning a lesson, reflecting on a lesson, discussing concepts with their peers, etc.). The topic in our case was R&P and lesson planning. The PMTs from the five participating countries have different mother tongues, but all of them learned English as a second language, so texts published in English were looked for. The teacher educators from the five institutions suggested texts that they thought appropriate for their PMTs. The whole team then met and agreed on the following list.

First, Silverman and Even's (2015) paper on modes of reasoning promoted in the tasks in mathematics textbooks was selected. It was deemed an easy and relatively short paper with a fairly basic theoretical part (which may not be understandable to an inexperienced reader). It introduces seven modes of reasoning based on concrete examples of tasks that are part of the curriculum in all five participating countries.

Second, Ellis et al.'s (2019) paper was used to introduce PMTs to possible teacher moves when implementing a mathematics task. The paper is rather long and has a lengthy theoretical part. However, to understand the table in which such moves are divided into four types (eliciting, responding to, facilitating, and extending student reasoning), it suffices to read about ten pages of the text, which includes many examples of such moves.

Third, to aid lesson planning, van den Akker's (2004) text introduced the PMTs to the "spider web" concept. Around the central concept of rationale for teaching (Why are students expected to learn this?), the "spider web" includes aspects of lesson planning the PMTs should consider (Aims or objectives, Content, Learning activities, Teacher and student role, Material and resources, Grouping, Location, Time, and Assessment).

Finally, two texts were selected to introduce PMTs to socio-mathematical norms. Hiebert et al.'s (1997, pp. 46–50) chapter presents the reader with the social aspect of mathematics learning in an easy-to-follow way through four guidelines which help create norms of a healthy social culture in mathematics lessons and which are part of the curriculum in all the participating countries ("Discussions are about methods and ideas", "Student choose their own methods and share them with others", "Mistakes are sites for learning", and "Correctness is determined by the logic of mathematics"). Stephan's (2020) encyclopaedia entry on social-mathematical norms was selected as an example of an up-to-date conceptualisation.

As the PMTs were to use the introduced concepts (teacher moves and parts of the "spider web") in their lesson planning (see Section 4.2.2), they needed to master their mother tongue equivalence. Thus, it was decided that the table with the moves and parts of the "spider web" would be translated into their language if the teacher educator deemed it necessary.

To prevent the inert knowledge problem and compartmentalisation of knowledge, the concepts introduced to the PMTs by the earlier texts were to be utilised in the on-site sessions in ways that would enable the PMTs to relate concepts to their future teaching more easily (see Section 4.3.1).

4.1.2 Lesson planning and lesson (re)design

Let us first present an overview of how the lesson planning process proceeded.

The diagram in Figure 4.1 depicts how the lesson plans (LPs) were developed. In Assignment 1, the PMTs were to select the task they considered an R&P task. No specific mathematical domain was prescribed; the choice of topic was entirely up to the PMTs. In Session 1, the tasks were discussed in the whole group. In Assignment 2, the PMTs started working in groups. They selected one of the tasks previously selected by the group members, or could choose a different one. Based on their reading of a text on teacher moves, they were to suggest teacher moves for implementing the task in a mathematics lesson. The tasks with the teacher moves were discussed in Session 2. In Assignment 3, the first version of the LP based on the task(s) previously elaborated was devised. They were discussed in Session 3, augmented as a homework task, and Version 2 of the plan was devised. Next, the plan Version 2 (one in each country) was enacted in a school, and this enactment was reflected upon in Session 4. Plan Version 3 was devised.

Effective lesson planning is at the heart of teachers' work, and including it in teacher education programmes is highly valued. The key role in planning and enacting mathematics lessons is played by mathematics tasks. Thus, we decided to include activities in which PMTs are encouraged to select and

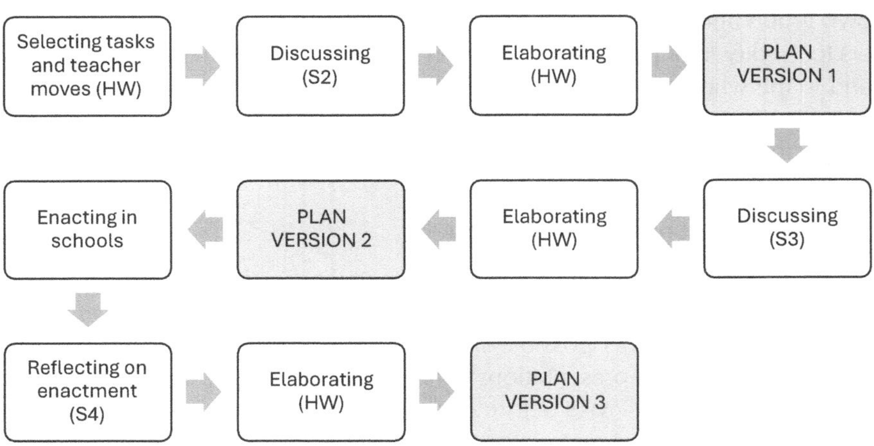

FIGURE 4.1 The origin of teaching plans (HW – homework, Sx – session number x).

modify tasks so that they could be included in an LP. Since pre-service teachers find it difficult to produce their own tasks and tend to rely on textbooks or Internet resources (Ulusoy & İncikabi, 2023), we asked PMTs to look into the existing curriculum resources for R&P tasks and to modify them rather than make their own. The designed tasks were to be the basis of an LP.

Cevikbas et al.'s (2024) review of studies concluded that the quality of LPs is enhanced if the teachers are asked to anticipate classroom situations such as students' strategies, questions, or difficulties. Thus, such prompts became part of the LP template provided to our PMTs in the Intervention. The lesson planning template was based on the earlier text (van den Akker, 2004) and the spider web so that the theoretical notions present in the spider web could be better connected to the act of teaching.

Another feature of the lesson planning activity in the Intervention was that it was organised in small groups of PMTs, enabling mutual feedback that potentially leads to more elaborate and thought-out plans.

Finally, lesson planning skills are enhanced by tailored feedback from experts and peers (Cevikbas et al., 2024) and thus, regular feedback from both parties was an ongoing feature of our Intervention.

4.1.3 Reflection on lesson plans and enacted lessons

Reflection, both self-reflection and collaborative reflection, is an important aspect of teachers' work. It enables them to critically examine their experiences and presumptions about teaching and improve their teaching practice. In our Intervention, we organised reflective activities representing common qualities of reflection (Ward & McCotter, 2004): reflection is situated in practice; it is cyclic in nature; and it makes use of multiple perspectives.

The PMTs were asked to reflect on the subsequent versions of LPs of their own group and the other groups and to reflect on the feedback from the others to modify their plans. This reflection was done both individually in writing and in the whole group during the sessions. A major reflective task consisted of reflecting on a video-recording of the enacted lesson, again individually and in the whole group.

4.2 Enactment of the Intervention

Figure 4.2 presents an overview of the Intervention, including the data collection types in each stage. To support the implementation of the Intervention in as similar a manner as possible in the five institutions, the team prepared common PowerPoint presentations guiding Sessions 1, 2, and 3 and common assignments given to PMTs before each session and following it (some are in the Appendices). These materials have been translated into local languages in CZ and SK. In IT, NO, and TR, only the English versions were used.

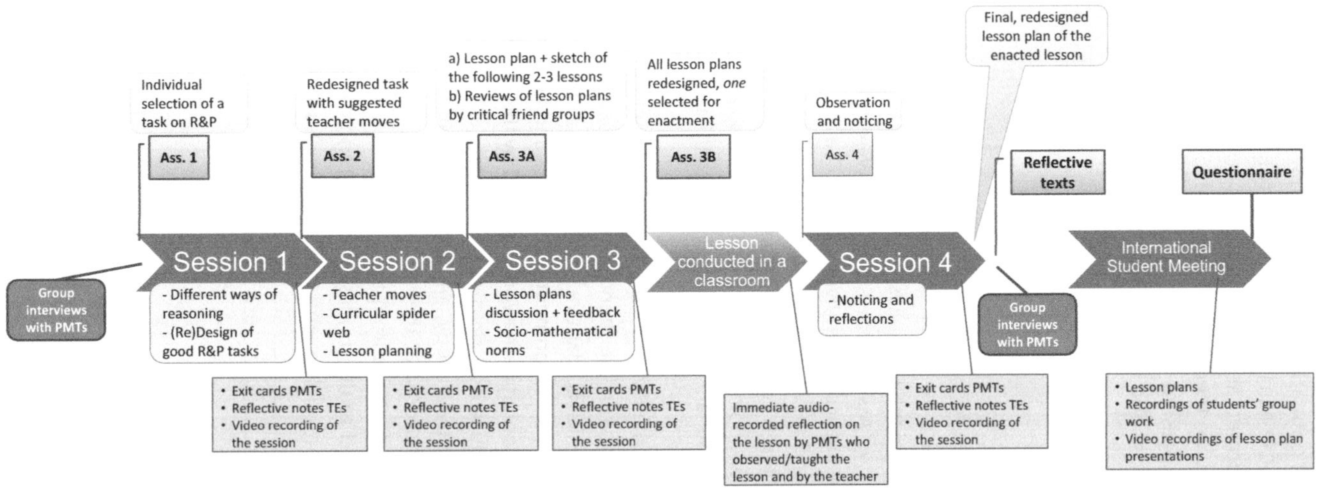

FIGURE 4.2 Overview of the intervention (Ass. = assignment).

Teacher educators who prepared the plan for the Intervention implemented it within their teacher education course, which was part of PMTs' programmes. Another member of the MaTeK project, a "critical friend", observed the sessions. They were informed about the goal and content of the Intervention, and their task was to provide feedback to the course leaders.

This section will present a common plan for the Intervention. The next section will briefly describe its enactment in each country.

SESSION 1

The goal of the first session was to acquaint the PMTs with basic modes of reasoning in mathematics for lower secondary pupils and how mathematical tasks and their implementation in lessons can promote them.

Before Session 1, in Assignment 1, the PMTs were asked to find the place of R&P in the national curriculum, individually select a mathematical task on R&P for Grade 8 or 9, and justify their choice. PMTs in the participating countries were educated to teach pupils of different grades, and grades 8 and 9 were found to be the overlapping ones. The PMTs were also asked to read Silverman and Even's (2015) paper and to relate their task to the modes of reasoning presented in the paper.

During Session 1, the PMTs discussed the goals of mathematical teaching. They were acquainted with five strands of mathematical proficiency (Kilpatrick et al., 2001), one of which is adaptive reasoning, which includes R&P. The PMTs shared their responses to Assignment 1 in groups and later in the whole class. Next, the ways of reasoning framework prepared by the MaTeK team (Sevinc et al., 2022) was introduced on prepared examples (e.g., appeal to authority, deduction, reasoning by analogy, and reasoning with empirical arguments). The PMTs then discussed what they understood as mathematical reasoning, and its five conceptualisations from the literature were presented. The rest of Session 1 was devoted to the (re)design of appropriate R&P tasks (e.g., tasks testing claims, justifying claims, and making claims) using examples of tasks.

SESSION 2

The goal of the second session was to develop the PMTs' insight into R&P tasks further, with a particular emphasis on teacher moves when implementing them in the lesson, and to introduce the PMTs to principles of designing teaching.

In Assignment 2 before Session 2, the PMTs were to read 11 pages of Ellis et al.'s (2019) paper on teacher moves in mathematics teaching and to select one R&P task for Grade 8 or 9 (they could draw on the tasks they selected in Assignment 1 and modified in Session 1), modify it if necessary, suggest concrete possible teacher moves for this task which could be used to support pupils' R&P, and to describe pupils' learning and their possible reactions to these moves. They were to complete this task in groups. Next, they were to read a summary of van den Akker's (2004) text on the "spider web" and think about how they would use it to plan teaching.

During Session 2, the discussion focused on Ellis et al.'s teacher moves in relation to the PMTs' responses to Assignment 2. Individual groups presented their work and discussed it with their peers. In the second part of Session 2, the discussion centred around the concept of levels of curriculum (supra, macro, meso, micro, and nano) and types of curricula (intended, implemented, and attained), and in the third part, the PMTs discussed individual parts of the "spider web" and how they could use it for planning a lesson.

SESSION 3

The goal of Session 3 was to develop further the PMTs' ability to design lessons based on R&P and to acquaint them with the concept of socio-mathematical norms.

In Assignment 3, before Session 3, the PMTs worked in the same groups as previously. With the help of the "spider web", they should elaborate a detailed LP with a sketch of the following two to three lessons. Based on the reading on socio-mathematical norms, they should connect it to their LPs. If possible, the PMTs were asked to comment on one LP of another group.

Session 3 was fully devoted to the PMTs' assignment. Each group presented its LP and discussed it with the others. The focus was on selected tasks, suggested teacher moves for implementing them in the lesson, pupils' possible reactions, mistakes, and problems, and socio-mathematical norms promoted in the LP. After Session 3, the PMTs were to change their LPs according to the feedback from their peers and teacher educators.

SESSION 4

The work leading to Session 4 was originally optional.[1] If an opportunity to implement the LP in the classroom arose, the PMTs or an in-service teacher cooperating with the department[2] were to do so. The lesson was video-recorded, and the recording was shared among the PMTs.

Before Session 4, the PMTs were asked to watch the lesson video and write a reflection based on prompts (e.g., the aims of the lesson and their fulfilments, R&P opportunities provided by the teacher, and how the pupils accepted them). Session 4 was fully devoted to the common reflection of the lesson video, using the PMTs' written reflections. Possible changes to the teaching plan were discussed. Next, the PMTs were asked to make the changes the group agreed on and submit the final version.

International student meeting

When the Interventions had finished in all the institutions, one LP per country, the one that was enacted in the class, was translated into English and sent to the other institutions. The plans were the focus of a two-day International Student Meeting in Prague. It aimed to advance the PMTs' lesson planning skills by making them aware of the cultural aspects of mathematics teaching. The PMTs first worked in national groups and presented and discussed the national LPs, selected aspects of the national curriculum, mathematics teaching, and future teachers' education with others. The core of the work consisted of on-site work in internationally mixed groups of PMTs, which jointly developed a lesson on a given curriculum topic area. Finally, the groups presented their plans and discussed them with the others in the plenary meeting.

4.2.1 Intervention per country

Table 4.1 shows the number of participants in each country. Except for Norway, the participants were not selected in any way. All the PMTs in the university course within which the Intervention was organised were involved. More information about the participants' background can be found in Chapter 3.

In **Czechia**, all the PMTs were master's students, future lower and upper secondary mathematics teachers (grades 6–13, age 11–19). Eight of them were full-time students, and 11 were part-time students.[3] All but one or two courses (which are in English) are in Czech in their programme. The Intervention was conducted within the third mathematics education course in the last (fifth) year of their university studies (three 45-minute sessions once a

TABLE 4.1 Teacher educators, critical friends, and the number of participants in each country

Place	*Czechia (CZ)*	*Italy (IT)*	*Norway (NO)*	*Slovakia (SK)*	*Türkiye (TR)*
Teacher educator	Nad'a Vondrová	Benedetto Di Paola	Iveta Kohanová	Mária Slavíčková	Didem Akyüz Işıl İşler Baykal[a]
Critical friends	Jakub Michal	Giovanna Rinchiusa Claudio Fazio	Birgit Pepin Melih Turgut	Tünde Kozánek Kiss Emília Miťková Peter Vankúš	
Number of PMTs	19	12	16	17	16

[a] In TR, both teacher educators taught the course, providing each other feedback.

week). Some students had experience teaching as unqualified teachers. They read some scientific papers in the two previous mathematics education courses, but only in Czech. The data collection tools (see Section 4.3) and part of the reading material in the Intervention were translated into Czech. PMTs submitted their assignments in Czech.

In **Italy**, all the participants[4] were students in the Master's degree programme in mathematics. The Intervention was conducted in Italian in their first or second (last) year of study, within a Teaching Methodology in Mathematics Education course. It was their first course on mathematics education. The PMTs had on-site sessions twice a week, five hours per week. They had little experience in teaching mathematics at school. The English reading and writing competencies were acceptable; thus, all the materials were assigned in English. PMTs submitted their work in Italian.

In **Norway**, all the PMTs were master's students in the Grades 5 to 10 programme. The Intervention was conducted in their fourth year of study, within a Learning and Teaching of Mathematics course, in which PMTs have classroom sessions twice a week, each lasting four times 45 minutes. In total, 42 students took the course. However, only 21 agreed to participate in the study.[5] Five of these did not participate in three sessions; thus, they were excluded from the study after all the data was collected. As the main course teacher is not Norwegian, the course was mainly taught in English, although the PMTs spoke Norwegian in the class. In Norway, it is a common practice to have almost all readings at the master's level in English, and most PMTs speak English well. Thus, all the Intervention materials were in English, and they submitted their work in Norwegian.

In **Slovakia**, all the PMTs were master's students in their fourth year of study to be mathematics teachers at secondary schools (grades 5–13, age range 11–19). The Intervention was held during their first semester in the master's studies and was included in the mathematics education (1) course, organised three times, 45 minutes per week. Similarly to the Czech Republic, some reading and data collection tools were translated into Slovak, and the assignments were submitted in Slovak.

In **Türkiye**, 12 participants were PMTs, while four already had an undergraduate degree (and thus are not, technically, PMTs in the Turkish context) and were enrolled in the Master's programme. Three PMTs were preparing to teach upper secondary school mathematics, while the remaining nine were enrolled in a programme to teach lower secondary school mathematics. One of the graduate students was an in-service teacher who had been teaching for around six years. The medium of instruction in METU is English, so the participants were expected to read academic texts in English; therefore, the materials were delivered in English. All assignments were submitted in English, except for exit cards and reflective texts, which could be written in Turkish.

4.2.2 Variations of enactment

In all the countries, the Intervention was implemented as planned (see Section 4.3.1) except for some departures from the plan caused by local circumstances. Table 4.2 presents chronologically those stages of the Intervention in which any variation occurred.

The Intervention was implemented at different times of the academic year in each country. The teacher educators (both the ones who taught the course and critical friends) regularly met online while the interventions were conducted and shared experiences using the shared tools, mostly documenting whether they had the required effect (e.g., whether the PMTs understood the assigned reading and how they responded to individual assignments). This enabled the others who had only started with the Intervention to use this experience to change the original design slightly to suit their needs better.

4.2.3 International student meeting

The last stage of the Intervention was the International Student Meeting, Prague, 18–20 May 2023. Thirty PMTs (7 from CZ, 10 from IT, 5 from NO, 5 from SK, and 3 from TR) and ten teacher educators (two from each country) participated in the meeting. The PMTs were grouped into seven groups,[6] each consisting of four or five PMTs of different nationalities (Table 4.3). These mixed groups were tasked with developing an LP related to geometry and measurement, using the spider web model, with a time allocation of 180 minutes. The only additional instruction provided was:

TABLE 4.2 Variations in the enactment of the intervention per country

	Czechia (CZ)	*Italy (IT)*	*Norway (NO)*	*Slovakia (SK)*	*Türkiye (TR)*
Session 1	3×45 mins for full-time PMTs; Session 1 and 2 combined for part-time PMTs into 3×45 mins	4×50 mins	4×45 mins	3×45 mins	3×50 mins; Activity 2 ("What is reasoning") moved to Session 2
Session 2	3×45 mins for full-time PMTs	5×50 mins	4×45 mins	3×45 mins	3×50 mins; Activity 2 ("Types of curriculum") skipped due to the lack of time
Session 3	3×45 mins for full-time PMTs; part-time PMTs watched a video-recording of Session 3 with full-time PMTs and had an extra task	4×50 mins	4×45 mins	3×45 mins	3×50 mins
Assignment 4		Not done due to ethical issues			Completed at the end of Session 4 due to ethical issues
Session 4	3×45 mins	5×50 mins, hybrid mode; PMTs watched the videos on site	4×45 mins	2×45 mins	3×50 mins; PMTs watched the video on site

TABLE 4.3 Nationalities of PMTs in the mixed groups

Group	*Nationalities of PMTs*	*Group*	*Nationalities of PMTs*
Group 1	CZ, IT, NO, SK	Group 5	CZ, IT, IT, NO, SK
Group 2	CZ, IT, NO, TR	Group 6	CZ, IT, IT, SK
Group 3	CZ, IT, IT, NO	Group 7	CZ, IT, SK, TR
Group 4	CZ, IT, NO, SK, TR		

> The lesson should support pupils' reasoning (and/or proving) and should be related to new knowledge, for example, area/volume of figures/solids, and should not focus on the Pythagorean theorem.

The topic of the Pythagorean theorem was excluded because its proof and related activities were well known to the PMTs. In the pre-interviews, most PMTs across all countries frequently mentioned this theorem as one of the first examples they thought of when asked about familiar R&P types or what aspects of R&P should be taught in schools. On the other hand, the topic of measurement was selected due to its complexity and well-known problems of pupils in it (e.g., memorising formulae rather than approaching the topic conceptually, Battista, 2007; Huang & Witz, 2011; Kordaki & Potari, 2002).

4.3 Data collection and analysis

Various data were collected to answer our research questions presented in Section 2.3. The violet rectangles in Figure 4.2 depict at which stage of the Intervention we collected them, and Table 4.4 shows how the data relate to the research questions. Some data were collected in the PMTs' local languages (Czech, Italian, Norwegian, Slovak, and Turkish). They were translated into English for analysis.

4.3.1 Data collection strategies and data analysis for RQ1

4.3.1.1 Data collection

Two data types were collected to answer RQ1: group interviews and a final individual reflective text. Before and after the Intervention, semi-structured (or "guide approach interview", Cohen et al., 2007, p. 353) group interviews with PMTs were conducted. The semi-structured nature of the interview enabled us to elaborate on issues and themes that emerged during the interview while keeping in mind the main aim. As the focus was not on PMTs on the

TABLE 4.4 Data per RQ

RQ	*RQ1 (beliefs)*	*RQ2 (LPs)*	*RQ3 (intervention features)*
Initial group interviews	■		
Final (group) interviews	■		■
Exit cards			■
Reflective text	■		■
Recordings of discussions in sessions		■	■
Assignments – versions of the LP		■	
LPs from the Student Meeting and recordings of group work and presentation of LPs		■	
Field notes		■	■
Questionnaire			■

individual level, we opted for group interviews, which can generate a wider range of responses than individual interviews (Cohen et al., 2007, p. 373). On the other hand, group interviews do not allow for determining the conceptions of individual PMTs, making it impossible to use quantitative analysis. In our study, it was left to the PMTs to make groups.

The pre-interviews conducted before Session 1 of the Intervention aimed to explore the PMTs' beliefs concerning teaching mathematics and, in particular, reasoning and proof, and their knowledge of the national curriculum and resources in terms of R&P. Specifically, the questions addressed to answer RQ1 included: *What forms of reasoning and proofs are you acquainted with? In your view, which forms should be included in the curriculum for grades 5-10/5-8? What is your perception of the role that reasoning and proof play in teaching lower secondary mathematics? How significant do you consider reasoning and proof to be in your instructional practice?* The post-interviews were conducted after Session 4 of the Intervention, after reflecting on the enacted LP. Thus, the questions included: *What have you learned about different modes/ways of reasoning and proof? How did it help you select/design a suitable R&P task? Did your beliefs regarding the role/importance of R&P in teaching evolve/change?*

During both pre- and post-interviews, the interviewer responded "dynamically" to the PMTs' statements and generated pertinent follow-up questions. The PMTs engaged with each other's responses, provided further elaboration, and posed inquiries independently. The duration of the interviews fluctuated

TABLE 4.5 Number of participants per country in the pre- and post-interviews

	Pre-interviews				*Post-interviews*		
	N	*Groups*	*Interviewer*	*Code (x – group number, y – PMT's number)*	*N*	*Groups*	*Interviewer*
CZ	19	6	TE	Cx.y	8	NA[b]	TE
IT	9	3	TE	Ix.y	6	3[a]	TE
NO	16	4	ST	Nx.y	7	3[a]	TE
SK	15	4	ST	Sx.y	14	4	ST
TR	16	6	TE	Tx.y	15	6	TE
	75				**50**		

[a] The composition of the groups differed from the pre-interviews, but all PMTs participated in the pre-interviews.

[b] In the case of CZ, individual interviews were conducted due to the problems getting PMTs' reactions after the end of the semester.

between 25 and 60 minutes, contingent upon the size of the group (comprising 2, 3, 4, or 5 PMTs) and the complexity and depth of the ideas and reflections articulated by the PMTs.

Table 4.5 presents the information about the participants involved in the interviews. The interviews were conducted by teacher educators overseeing the Intervention (TE) or by advanced Master's students, who were enlisted to work collaboratively with the local MaTeK team (ST).[7] Some interviews were conducted online (CZ, partially IT, NO, and TR in the post-interviews), while others were in person. The number of post-interviews is smaller as not all the PMTs were available after the end of the semester.

4.3.1.2 Analysis

Audio or video recordings of the interviews were made and later transcribed for the analysis. The transcripts underwent thorough analysis employing qualitative methodologies as delineated by Saldaña (2015), utilising MAXQDA software. First, a coding of pre-interview transcriptions was executed through a combination of inductive and deductive approaches. The preliminary coding framework, grounded in pertinent literature related to beliefs and R&P, such as the works of De Villiers (1990), Hanna (1990), Harel and Sowder (1998), and Knuth (2002), was expanded with additional codes that emerged throughout the interview data analysis. Upon completion of the coding process, the coded passages were subjected to iterative readings to discern overarching themes and contexts. Consequently, categories were delineated:

Notion of R&P, *Roles of R&P*, and *Concerns and difficulties associated with the instruction of R&P*.

An example of coding is the statement, "... in those classes or in the way I teach, there's just not always as much time for it [R&P] as there needs to be" (code *Time Limitations*, category *Concerns and Difficulties*). The categories sometimes overlap. For example, the statement "... you could say that, like actually, his process was a proof, yeah, that like he convinced the others..." was coded under the category of *Role of R&P* and *Notion of R&P*.

The initial coding was conducted on the Czech data, and the resulting coding framework was shared with the other countries (SK, NO, IT, and TR). The framework was slightly modified to suit the needs of all the data.

The analysis of post-interview transcriptions was done differently. The transcriptions were repeatedly read in the light of the three main categories determined for the pre-interviews, and any new ideas were noted.

At the end of the Intervention, the PMTs wrote an individual reflective text (the final assignment in Figure 4.2) in which they were asked to reflect on the whole Intervention in terms of how they perceived its content (e.g., what they learned, how the group work helped, how confident they felt about planning a mathematics lesson focused on R&P, see Appendix 2). Not all the PMTs submitted their final assignments (Table 4.6 in Section 4.3.3).

The final assignments were used as complementary data only, as their primary goal was not the PMTs' conceptions of R&P but their perception of the Intervention as a whole. They were searched for the mention of the three categories identified earlier in the analysis of the transcriptions of interviews.

4.3.2 Data collection strategies and data analysis for RQ2

4.3.2.1 Data collection

The Intervention's core consisted of preparing LPs focused on R&P. The assignments (see Figure 4.2) gradually led the PMTs from a single task towards a full-blown LP around the task(s) based on a spider web template. These assignments provided the core data for answering RQ2.

Towards the end of the Intervention, the PMTs decided which LP they would enact in the classroom and reflect on in Session 4. It was this plan that was submitted to the "international round" to be presented at the Student Meeting in Prague. This plan, respectively its three versions (Version 1, 2, and 3),[8] were used in our analysis towards answering RQ2.

Finally, lesson planning was one of the activities during the Student Meeting. It resulted in mixed international groups of PMTs preparing and presenting one LP each. These plans were also used to answer RQ2.

Two types of analyses were carried out: (1) analysis of LPs; and (2) development of LPs over time.

4.3.2.2 Analysis of lesson plans

First, Version 3 of one plan (the revised version of the enacted lesson) per country was analysed. For CZ, NO, and TR, Version 3 of the national plans concerns one lesson (40 or 45 minutes long) and includes brief information about the following lesson(s). The plan for SK concerns a double lesson (90 minutes); IT includes five lessons (60 minutes each). In both cases, the analysis concerns all the planned lessons. One researcher translated the plans into English, and the other checked the accuracy of the translation.

Document analysis was employed as the primary method to examine the LPs developed by PMTs from each country. Document analysis involves examining and interpreting data to elicit meaning, gain understanding, and develop empirical knowledge (Corbin & Strauss, 2008). This method allowed us to explore the LPs in depth and to understand the underlying intentions, pedagogical strategies, and educational priorities of the PMTs within their respective national and cultural contexts. In the analysis, we attended to aspects of the curricular spider web (van den Akker, 2013), and we also examined the core mathematical content and pedagogical strategies, focusing on how they facilitate R&P.

Document analysis is an iterative process that involves reading and interpreting documents and combining elements of content analysis and thematic analysis (Bowen, 2009). This method allowed us to organise information from the LPs into categories aligned with our research question. We began by creating a table with 11 categories based on the curricular spider web model (van den Akker, 2013), which outlines core aspects of educational design, such as objectives, instructional strategies, assessment methods, and teacher and student roles. This framework provided a structured way to evaluate the pedagogical intentions and practices embedded in each LP. Each LP was reviewed separately by two pairs of researchers. During this phase, we thoroughly read the LPs to complete the initial table, noting relevant information within each of the 11 categories.

This first analysis aimed to capture each plan's basic structure, key elements, and unique characteristics. After completing the table, the findings were compared to ensure consistency, reliability, and accuracy. The two pairs of researchers discussed any discrepancies and refined their observations to reach a consensus on each LP's key features. Subsequently, a preliminary summary of each LP was collaboratively drafted to outline the LP's distinctive characteristics.

In the next round of analysis, we focused specifically on R&P within each LP. This involved a more detailed reading and interpretation of the text, examining how the LP was structured to facilitate reasoning, the types of reasoning explicitly or implicitly mentioned (e.g., empirical reasoning, deductive reasoning), and the conceptualisation of reasoning (Herbert et al., 2015), as outlined in the plan.

We also analysed (implicitly or explicitly stated) anticipated *paths of reasoning*, that is, the transitions between different types of reasoning that pupils are expected to make within the lesson. For example, some LPs anticipated pupils progressing from empirical reasoning to deductive reasoning under the teacher's guidance.

Finally, insights from the table and the R&P analysis were synthesised to produce a comprehensive understanding of each LP. We identified patterns and themes across the LPs, enabling a comparative analysis of different educational approaches. Through this thematic analysis, we explored similarities, such as the inclusion of group work and whole-class discussions across all plans, as well as differences. For instance, the Italian LP showed a juxtaposition of Euclidean and non-Euclidean geometry to explore distinct proof schemes, while the Turkish plan provided the most detailed resources and assessment methods in R&P. This synthesis enabled us to articulate how PMTs conceptualise mathematical reasoning and facilitate its development within their instructional designs, reflecting cultural distinctions within and across educational contexts.

Next, LPs designed in mixed international groups during the Student Meeting were analysed similarly to the national LPs. These plans were already in English, which was the language common to all the PMTs working in groups. As complementary data, we also used videos of the mixed group presentations of their LPs at the end of the Student Meeting. These presentations provided a more comprehensive and nuanced understanding of how the PMTs envisioned the lessons. However, not all LPs were fully completed due to the limited time available to the PMTs during the meeting in Prague.

4.3.2.3 Development from Version 1 to Version 3 of national lesson plans

As already stated, there are three versions of each plan (see Figure 4.1). Versions 1 and 2 were elaborated both in homework assignments and during discussions in the sessions. Version 2 of one LP in each country was then enacted in the class and reflected on during Session 4. Based on this reflection and the enactment, the group elaborated the plan into Version 3. Our analysis focuses on the changes or modifications between subsequent versions of the plan and their possible causes.

The main data was complemented by teacher educators' field notes from the sessions, written feedback the PMTs were asked to provide (at least one for each plan), and video recordings of sessions (Sessions 3 and 4). In some cases, teacher educators also provided written feedback for the plans.

As described earlier, document analysis methods were employed. The appropriate documents (Version 1 versus Version 2 and Version 2 versus Version 3) were compared to analyse changes between plan versions. This comparison was facilitated by the fact that the documents were structured in the same template with categories corresponding to the "spider web": Aims and objectives, Content, Learning activities, Teacher and student role, Material and resources, Grouping, Location, Time, Assessment, Homework, and outline of the next lessons.

The researchers from the appropriate countries conducted the analysis of the Version 1 versus Version 2 plan.[9] The two documents were put side by side, and the content of the identified categories was compared (the category Homework was merged with the category Learning activities, and the category Outline of the next lessons was dispersed into the earlier categories). Next, it was determined what the likely cause of such a change was: whether it was a response to (a) the teacher educator's feedback in Session 3, (b) the discussion in Session 3, or (c) the discussion within the group of authors of the plan after Session 3. The complementary data (the video-recording of Session 3 and our field notes) were used towards this goal. When no evidence could be found of cause (a) (teacher educator) or (b) (peers), cause (c) was assigned. The results of this analysis were filled in tables. An example of such a table is in Table 5.2 in Section 5.2.3.

The analysis of the Version 2 versus Version 3 plan was organised differently, as Version 3 was translated into English for the Student Meeting in Prague. The researchers from each country augmented the document depicting Version 3 with changes from Version 2. Namely, any text not present in Version 2 and added in Version 3 was marked red, and any text deleted from Version 2 was added in the document on Version 3 as strikethrough text. An example of a part of this document for the Czech plan is in Figure 4.3.

The unit of analysis was one meaningful modification. Each unit was coded using one category of the "spider web" described earlier. Two coders independently coded the modifications of all the plans and then met to negotiate

Aims and objectives (linking to the curriculum if possible): *What are the expected learning goals of the pupil?*
At the end of the lesson, pupils will be able to ~~(abilities)~~: - decide and **justify the correctness of the proposed solutions** to the tasks on the worksheet - correct incorrect solutions from the worksheet (justify the incorrect reduction in fractions) - **visually or verbally interpret algebraic expressions** from the worksheet ~~- define what "proof" means~~ ~~At the end of the lesson, pupils will be able to (concepts):~~ ~~- relationships for the squares of addition and subtraction~~ ~~- relationship "the difference of squares"~~

FIGURE 4.3 Illustration of part of the document comparing Versions 2 and 3 of the Czech plan.

TABLE 4.6 Part of the document with the coding of changes between Versions 2 and 3 of the Czech lesson plans

Country/ Change	*Code*	*Short description*	*Spider web*	*Nature*	*Possible cause*	*Type*
CZ	CZ02	Specified aims and objectives	Aims and objectives	They decided not to use the distinction between abilities and concepts as they overlapped, and it was not easy to decide what goes where. They focused the goals on specific tasks (earlier they were general).	In Session 2, the educator asked if PMTs felt the formulated goals were fulfilled. PMTs reacted that they would not be able to measure how the goals were fulfilled as they were too general. The educator suggested specifying them and PMTs agreed.	Making what was originally planned more explicit.

any differences. During this process, it was noted that there were two main types of modifications made by the PMTs: those that led to a *change* in the LP (such as adding a new task to the plan) and those that only made what was originally planned more *specific* (*explicit*). Thus, each modification was also coded "change" or "specification". Finally, for each modification that led to a change in the LP, the teacher educator from the respective country was asked whether they could determine the cause of such a change based on the complementary data from the whole Intervention. An example of the results of this process is in Table 4.6.

4.3.3 Data collection strategies and data analysis for RQ3

4.3.3.1 Data collection

Three types of data were collected to investigate how the PMTs perceived the design features of the Intervention.

First, the PMTs were asked to fill in *exit cards* after each session to reflect on their learning (see Appendix 4). The exit cards were distributed on paper. The PMTs were to describe their overall feelings, say what important things they learned, which part of the session they liked (and why), and what was

TABLE 4.7 Final assignments and exit cards submitted by the PMTs

	Total	*Number of final assignments*	*Number of exit cards*
CZ	19	4 (+7)[a]	32
IT	12	7	43
NO	16	10	48
SK	14	12	55
TR	16	15	44

[a] As the number of assignments submitted by the Czech PMTs was very low, we reached some of them again after the end of the semester and managed to conduct individual interviews with seven PMTs. The interviews had the same focus as the reflective text. In the analysis, the data from the interviews were treated like the data from the final assignments.

difficult for them. Not all the PMTs submitted their exit cards (Table 4.6). For Czech PMTs, the idea of exit cards was new, and judging from their responses on site, they did not see much sense in completing them. They mostly replied using one-sentence answers and only rarely provided more feedback or evaluated their learning.

Second, the *Final Assignment* after the fourth session provided data for RQ3 (see Appendix 2 for the task given to the PMTs). Again, not all the PMTs submitted their final assignment, as it was assigned after the end of the semester (Table 4.7).

Finally, the *Questionnaire* at the end of the Student Meeting also provided data for RQ3. The participants were asked to share their reflections on the most valuable aspects of the programme, the key insights they gained, also in terms of lesson planning with a focus on R&P, and their experiences working in international groups. All 30 participants responded to the questionnaire, with the option to answer in English or their mother tongue. For analysis, all responses were then translated into English. The questionnaire is provided in Appendix 3.

4.3.3.2 Analysis

The data for RQ3 were analysed qualitatively, again using document analysis procedures. The analytical units were individual responses related to one theme. First, the PMTs' responses in the exit cards were analysed in terms of their evaluative comments on the activities realised within the session in question to gain insight into how individual activities in the Intervention worked and whether they reached the required aim. Such comments were then merged into three categories corresponding roughly to the design features upon which the Intervention was built (Reading, Task design and lesson planning, and Reflection, see Section 4.2).

Next, the same was repeated for the data from the reflective text. Only comments that conveyed what the PMTs learned (e.g., in terms of R&P and lesson planning) and were related to the design features of the Intervention were selected. For example, the statement that a PMT understood R&P better, which did not include what enabled such an understanding, was excluded from the analysis for RQ3. The frequency of statements was not followed, as we aimed to uncover emerging themes and not how prominent they were.

Similarly, questionnaire responses were analysed based on PMTs' evaluative comments on activities conducted during the International Student Meeting. The focus was on how and why these activities contributed to their learning or led to a shift in perspective or beliefs regarding lesson planning and/or the teaching of R&P. Additionally, we examined the challenges mentioned by PMTs to provide a more comprehensive understanding of the impact of the international collaboration.

Notes

1 But there were favourable conditions, and all the participating countries implemented this part of the Intervention.
2 In such a case, the teacher was given the whole LP and could ask questions about it before teaching it.
3 Part-time students study the same courses, but they have about two thirds of the on-site lessons that full-time students have.
4 There were originally 15 students in the course, but three of them did not complete it.
5 The PMTs who did not agree to participate in the study were present during the sessions and received the same teaching. They were not recorded, and their work was not analysed.
6 The division of groups was done by teacher educators so that there was a balance in terms of countries and gender.
7 Student teachers trained by the course leader conducted interviews. Familiarity with the interviewers, who were fellow students, was expected to encourage honesty, unlike interviews by the course leader. The interviews also supported the interviewers' master's thesis on changes in PMTs' knowledge for teaching R&P in the MaTeK project.
8 There were several groups in each country which prepared several LPs. It would be impossible to conduct an analysis for all of them, so it was decided to use one plan per country. See Figure 4.1 for a graphic way of how Version 3 of the LP originated during the Intervention.
9 For the CZ group, the situation was different from the other countries. The group members, whose plan was later selected for the international round, completely changed the content of their plan after preparing their first ideas. During Session 2, when presenting their ideas, the feedback from their peers was such that they did not lead to reasoning and proving, and both the peers and the teacher educator were sceptical that they could be elaborated into an R&P lesson. Thus, there is no Version 1 for this particular group of PMTs, and for the Czech group, only the changes from Version 2 to Version 3 were analysed.

5
RESULTS

In this chapter, we present our findings related to each of the three research questions.

RQ1: What are pre-service mathematics teachers' (PMTs)' beliefs concerning reasoning and proving (R&P) in teaching before and after the Intervention, and how can we understand these, considering that PMTs worked in five different (national and cultural) contexts?

RQ2: What are the characteristics of, and the similarities and differences among, the different lesson plans? In which ways did lesson plans change/progress during the Intervention, and how was PMTs' progress influenced by the set-up of the Intervention?

RQ3: How did the PMTs experience the Intervention?

5.1 PMTs' beliefs concerning reasoning and proving in teaching before and after the Intervention

Group interviews and final written reflections were used to understand the PMTs' beliefs concerning R&P in teaching before and after the Intervention. The analysis of the data from the pre-interviews revealed three categories: (1) PMTs' R&P concepts; (2) their perception of the role of R&P in the teaching and learning processes; and (3) their concerns about R&P in schools. The following pre- and post-Intervention results are structured according to these three categories, followed by a description of the category that identified how PMTs viewed the position of R&P within school mathematics in the post-interview, and the final reflections.

DOI: 10.4324/9781003533870-5

5.1.1 *The notion of reasoning and proving*

It is perhaps unsurprising that before the Intervention, the PMTs lacked clarity in defining their perceptions of R&P. While all except for Norwegian PMTs had ample experience with proving from university classes, they typically had no specific instruction on teaching R&P in their methods or mathematics education courses at the university. Rather than explicitly defining R&P, the PMTs referred to different roles of R&P in teaching, and from such statements, their conceptualisations of R&P could be implicitly derived. Three conceptualisations were more frequent among the PMTs across countries: (a) R&P as justification of correctness; (b) R&P as deduction of new pieces of knowledge; and (c) R&P as the description of steps.

(a) Reasoning and proving as justification of correctness

Before the Intervention, and partly after it, PMTs in all the country groups predominantly perceived R&P as a justification of correctness. The PMTs referred to pupils' capability to **substantiate the validity of their solutions**:

> ... to be able to, like, confirm, their [pupils'] solution is correct
>
> *(CZ-pre)*

> [...] to understand why and how what I did was correct. I think if you know why what you did was correct, you learn more than just having the answer
>
> *(NO-pre)*

PMTs stressed that reasoning "is not just describing what you've done" (NO-post); it goes beyond that and answers the Why question: "[the pupils] shouldn't immediately say the answer, but they should say why they did this" (SK-post).

Notably, even before the Intervention, and more so after it, some PMTs posited that R&P involves **making and seeing connections** within mathematics. According to PMTs, when pupils make connections, they integrate all their knowledge and can "see the big picture" (NO-pre); R&P enables pupils to make sense of mathematics:

> The child basically, for example, knows the topics. Then a new situation comes up, for example. By synthesising what they already know, they can solve that new situation, which could be a new problem, a new topic. They can create a new template in their head based on the previous ones. This seems like reasoning.
>
> *(TR-pre)*

> And just so that they [pupils] can perhaps even realise that the math is not just for me to calculate the entire workbook simply. But that they simply see some connections.
>
> *(SK-pre)*

> [R&P refers to] the process of constructing mathematical thinking and being able to draw coherent, well-founded conclusions.
>
> *(IT-post)*

R&P as the **justification of a theorem or procedure** was very frequent among PMTs:

> Why does multiplication work this way? Or why is pi what it is?[1]
>
> *(NO-pre)*

> For example, let's say we are investigating the interior angles of a pentagon… Since the sum of the interior angles of the quadrilateral and the sum of the interior angles of the triangle are generally known, maybe from there, you know, from the triangles, they can say, "There are three triangles in this. Three times 180".
>
> *(TR-pre)*

Some Italian and Turkish PMTs clearly distinguished R&P. Reasoning for them was something "informal" ("a chain of useful conclusions to arrive at the thesis", IT-pre) as opposed to "real" or "formal" proving:

> [Proof is a] process of deduction which, starting from premises assumed to be valid (called axioms) or from propositions already proved, determines the validity of a new proposition. Because of the formal correctness of the reasoning used.
>
> *(IT-pre)*

Overall, before the Intervention, the Italian PMTs emphasised the proof part of R&P more than the reasoning part. They contended that proof was important in later grades and mostly associated it with geometry as "the most appropriate place to prove" (IT-pre):

> Lower secondary school pupils do not need to learn to prove as we do [referring to formal proof], they can stop at an earlier stage. Upper secondary school pupils, however, should. They should be required to give rigorous proof (at least for some theorems).
>
> *(IT-pre)*

Almost all Italian PMTs highlighted a significant "big leap" (IT-pre) in the transition from "doing school and doing university mathematics" (IT-pre), indicating an educational phase wherein the function of R&P undergoes even more substantial transformation than within the school context:

> In school, you don't really prove, in my opinion, you learn the steps by heart, you try to understand the why of these steps, but you are not always really aware of the whole proving process. A lot of things are true without anyone proving them. But then you go to university, to math, for example, and everything becomes very rigorous. The need to prove arises.
>
> *(IT-pre)*

After the Intervention, some PMTs explicitly mentioned that their conceptions had changed or evolved. For example, they said that they "started thinking a lot more about it" (NO-post); "from now on, I'm going to approach the subjects that I'm going to teach and explain by thinking about the end, by thinking about where it came from... So, my ideas have changed too" (TR-post). By far the most prominent change in the PMTs' beliefs concerning R&P following the Intervention is apparent in their no longer seeing R&P only as rigorous proof. They stated that the Intervention made them realise that it is possible to use such R&P that is not purely formal or "entirely correct": "So, I became aware of this consideration. That it is possible to do [R&P] even partially [even] if it cannot be done properly." (CZ-post) The change was caused by being acquainted with various ways of implementing R&P activities during the Intervention:

> Before, I had the idea that only such a robust, deductive argument was okay to take as an argument and that during this, I somehow realised that on some level, pupils do not have a chance to make such an argument and that even if they use something other, it can be accepted, it's actually okay.
>
> *(SK-post)*

> Previously, I would have probably expected it to be just about proofs, you know, like pure proofs of whatever, which is kind of... not controversial, but sometimes it doesn't quite fit or there's no time for it at the primary level. But now I see it a bit differently – I perceive it as something that can be part of an everyday math lesson.
>
> *(CZ-post)*

> It sounds like a paradox, but we, as future math teachers, are more "ready" for formal proof than non-formal ones. At the university, we learn more about it but ... we realised that R&P is also something else! We didn't

> reflect on it before this intervention. Its role at school is something different from the formal one, ... It is something deeper. Thanks to this course, I started thinking a lot more about it.
>
> *(IT-post)*

(b) Reasoning and proving as a deduction of new pieces of knowledge

Before the Intervention, some PMTs saw R&P as the process of **deriving** (e.g., formulas and procedures) at the onset of new subject matter, which they contrasted with a simple articulation of such a formula or procedure.

> ... that initial derivation of, for example, the formula for the square of a binomial or [...] the sum and the difference [of binomials], [deduced by pupils by drawing squares]. Divide it [the square] so that the pupils can figure it out on their own and try to formulate it afterwards.
>
> *(CZ-pre)*

> You know, you could calculate the circumferences of different circles and prove [finding out] pi... experimentally.
>
> *(TR-pre)*[2]

> ...they don't remember formulas, but they might know the method or the path to calculate this or derive the given formula. This can also positively develop pupils' logical thinking.
>
> *(SK-pre)*

(c) Reasoning and proving as the description of steps

Before the Intervention, some PMTs characterised reasoning as describing the procedural steps that learners must undertake to arrive at a solution. According to their views, it simply involved enumerating the sequence of actions without providing justifications for those actions:

> I always want them to describe everything, too, how they reached that [solution] in the first place. But it's more that I want to justify why they reached the result this way and not that way, not why it is like that.
>
> *(CZ-pre)*

> In front of the board, presenting a solution and reasoning that if this equals that because I add these two together... Commented solution. Utilising some basic facts.
>
> *(SK-pre)*

Notably, after the Intervention, some PMTs highlighted that R&P is not a mere description of steps (as at least some had seen it previously):

> I better understand that reasoning is not just describing what you've done (as in many schools), but that it goes beyond that.
>
> *(IT-post)*

> If there was why they could solve it like that, then OK, we have elements of justification, but only describing how they solved the equation in words, that it is not, I think, R&P.
>
> *(CZ-post)*

5.1.2 The role of reasoning and proving in the teaching-learning process

Even before the Intervention, the PMTs overwhelmingly agreed that R&P was important, openly or implicitly, by pointing out its benefits and providing various explanations. Five roles emerged in the pre-interviews, and all appeared in the post-interviews, but this time the PMTs spoke about them with more conviction. The five roles were: (a) R&P serves to justify the existence and truth of mathematical knowledge; (b) R&P leads to pupils' engagement with mathematics; (c) R&P leads to a deep understanding; (d) R&P convinces pupils that something holds; and (e) R&P helps to communicate mathematics.

(a) Reasoning and proving serves to justify the existence and truth of mathematical knowledge

A significant role was attributed to R&P by PMTs in demonstrating that novel mathematical knowledge is grounded in rationale and does not arbitrarily originate, thereby fostering a more profound understanding among pupils regarding the structure of mathematics and the genesis of knowledge within the mathematics discipline.

> It seems to me that it [R&P] has like a big role in the sense of – especially reasoning – so that pupils can see, even if they might not understand it, that the proofs or the mathematical theorems don't just fall out of the sky, or that they're not just made up by some people somewhere who wrote it in a book, but that they're actually things that can be deduced.
>
> *(CZ-pre)*

> I will ask a rhetorical question: what is mathematics about, if not about this, about that reasoning? Mathematics as a science is exactly about that.

> This leads to this, and this leads to that. And that's what the kids should take away from math classes.
>
> *(SK-pre)*

Notably, in the post-interviews, some Czech PMTs mentioned the ideas of systematisation and connection between ideas, building on how mathematical knowledge originated in history. They considered it important to convey to their pupils that mathematics "didn't emerge in a day" and was not invented "by one person", but rather that ideas were interconnected, built upon one another, and derived from previously established concepts or statements. They believed R&P was an appropriate tool for achieving this.

> The proof is nice in the sense that it gives you some order in the mathematical structure. Basically, it's clear – we're gradually building up the mathematics, and it's kind of illuminating, you know, like transparent and clear. But the moment we just give them a formula, they don't know what to connect it with or how it's linked to other things.
>
> *(CZ-post)*

> And I see another positive aspect in it – that this is real mathematics. That it's the science itself, a genuine demonstration of how a mathematician actually works in their everyday practice. They're not just folding papers; they simply have a pencil, and a piece of paper, and they're using a certain kind of language.
>
> *(CZ-post)*

(b) Reasoning and proving leads to pupils' engagement with mathematics

Czech PMTs connected R&P with pupils' engagement with the subject matter and their comprehension and motivation by pointing to a detrimental effect of the perceived deficiency of R&P activities within the Czech curricula.

> I guess I didn't enjoy the math because none had done any deducing [new knowledge] for me.
>
> *(CZ-pre)*

> At the secondary grammar school, they just told us, this is how it is – this is how you solve it. There was no understanding there.
>
> *(CZ-pre)*

Similarly, a Turkish PMT and an Italian PMT believed that when R&P was used for deducing pieces of new knowledge, it could motivate pupils:

> The interest of the child…increases because it is also their own effort. You know, because they discovered it, instead of us telling it, they embrace it. They own that information. I think this is also a very important point.
>
> *(TR-pre)*

> In school, especially in secondary school, R&P should play an important role because pupils should be able to appreciate the beauty of mathematical proof and the reasoning that is useful to get there in the theorems that are studied in school, of course.
>
> *(IT-pre)*

(c) Reasoning and proving leads to a deep understanding

The view of teaching incorporating R&P as something that "explains", "shows pupils why it is so", or "leads them to deeper understanding" was common among PMTs even before the Intervention. This role was associated with their belief that exposure to R&P when learning something new would convey deeper comprehension and prevent the emergence of knowledge without understanding. Many PMTs linked R&P to helping pupils grasp why mathematical concepts are as they are, rather than simply memorising formulas:

> Proof should help pupils understand why a mathematical statement is true, rather than just verifying that it is true using a formula.
>
> *(IT-pre)*

> Have an understanding of why. Why do we learn this, first of all, and why, why are things the way they are? Why can we just … why does multiplication work this way? […] I think that it is a large part of the reasoning, and if you can explain why what you have done is right, when you have learned from it, then you have … proved it in a way, perhaps not a deductive proof, but yes.
>
> *(NO-pre)*

> I think it was Ali Nesin if I remember correctly. In a video on YouTube, he said, 'I don't memorise anything because my memory is very bad. But since I know the logic of everything, I can find it if I forget.' The first thing that comes to my mind with this example is that, of course, we don't want information as a pill; we actually want to learn its logic. And this part of learning its logic is precisely this method of understanding why and reasoning…I mean, it is something that saves us from rote memorisation.
>
> *(TR-pre)*

Some PMTs stressed that reaching a deep understanding through R&P would help pupils retain knowledge and help pupils become independent learners, which, however, was not seen as the role that school teaching of mathematics would necessarily fulfil:

> […] when we learn the proof of something, that is, when we learn where it came from, the information stays in our brains.
>
> *(TR-pre)*

> Often, however, things are taught by rote, without understanding why, just by writing down a sequence of steps that have been taken. The R&P role, therefore, in many cases, unfortunately, is a writing exercise, "only" a procedure description, without any real meaning. Much later, the true R&P role is understood.
>
> *(IT-pre)*

The perception of R&P as explaining and deepening mathematics and helping pupils move beyond procedural understanding (knowing how to apply algorithms) to conceptual understanding (knowing why algorithms work) prevailed at the end of the Intervention across all country groups. Concrete benefits stemming from a deep understanding were mentioned, such as more efficient problem-solving abilities and lower demands on pupils' memory. Some PMTs stressed that they realised this during the Intervention.

> My impression when I started with this course was that proof and reasoning were on the same level as the rest of the topics in mathematics …But then … I started to get the impression that this is important for creating a deeper understanding of mathematics, not just that they [pupils] know it, but that they understand why algorithms work.
>
> *(NO-post)*

> Pupils will then see into it a bit more and will also have conceptual understanding, rather than just following the idea that there is some formula to memorise without knowing where it came from or how it relates to the topic.
>
> *(CZ-post)*

Notably, in the post-interviews, the Czech PMTs widened the role of R&P in reaching a deep understanding of mathematics by saying that R&P had benefits for pupils in their everyday life, outside of the realm of mathematics. For example, "I find the overlap with everyday life important. Proof, in my opinion, develops the ability to think logically in areas beyond just mathematics." (CZ-post) Another Czech PMT expressed the same idea in broader terms, noting that understanding that things had their reasons was important even when solving everyday problems, where individuals could not simply accept everything they read or hear. Another benefit of R&P, extending beyond the domain of school mathematics, highlighted by the PMTs, was the development of communication skills: "I imagine R&P as a way to guide pupils toward perhaps better ways of expressing themselves." (CZ-post)

(d) Reasoning and proving convinces pupils that something holds

Before the Intervention, the role of proof as conviction in relation to convincing classmates or pupils in the class was only rarely mentioned:

> [R&P serves to] convince those classmates that I was right [...] even though most of them agreed on something, one individual was able to justify it and just defend their solution.
>
> *(CZ-pre)*

> To convince themselves or to convince each other [pupils] that something is correct or true.
>
> *(SK-pre)*

Even after the Intervention, this role was explicitly mentioned only by some Czech PMTs, who stressed the role of convincing oneself or others (their peers and the teacher) using facts: "in proving, it is not about being able to express oneself or about having a better opinion" (CZ-post). They also mentioned proofs using manipulatives as being particularly convincing for pupils.

(e) Reasoning and proving helps to communicate mathematics

Some PMTs' statements in the pre-interviews could be linked to the communication role of R&P ("including non-formal communication", I1.4-pre) in situations in which pupils communicated their thinking process while solving problems:

> The kids were describing how they actually came up with the solution to the problem, so when anyone takes the problem and sees the solution, they're supposed to understand how the kid solved it.
>
> *(CZ-pre)*

Two Norwegian PMTs also stressed the importance of pupils' language and symbolism development in connection with R&P, noting that pupils needed the tools to express their reasoning effectively. "You have to give them the tools they need to express it" (NO-pre) so that pupils will "be able to use it to explain why they arrived at the answer" (NO-pre).

Some Czech and Norwegian PMTs highlighted the importance of R&P in revealing pupils' thought processes. In that case, the teacher would be able to assess whether pupils understood the subject matter and discover potential misconceptions in pupils' thinking:

> So, for me, [R&P] definitely has a crucial [role] because to pick up on those pupils' thought processes and the way they think, I definitely need to know how they got to that particular result.
>
> *(CZ-pre)*

> And then you, as a teacher, also, in a way, get to know what they were thinking. So if you get a little into their heads, [...] if there is someone who thinks wrongly about something, then maybe you can discover what it is that they think wrongly.
>
> *(NO-pre)*

5.1.3 Concerns associated with teaching reasoning and proving

Five sources of concerns associated with teaching R&P to primary and lower secondary pupils were mentioned by the PMTs across countries in the pre-interviews: (a) curriculum; (b) time; (c) pupils' difficulties; (d) establishment of appropriate socio-mathematical norms in classes; and (e) PMTs' lack of prior experience and pedagogical content knowledge. These concerns were still present after the Intervention, but the PMTs conceptualised them differently.

(a) Curriculum

Although the numerous advantages of R&P activities outlined previously were acknowledged, the PMTs expressed apprehension regarding their ability to effectively deliver the mandatory curriculum within the time scope allocated to mathematics. For example, the Czech PMTs, who regarded R&P as instrumental in fostering conceptual understanding, viewed the national entrance examinations for upper secondary schools as an impediment, as these examinations, according to them, prioritised procedural skills over conceptual comprehension.

> But it seems to me that in the system we have now, that [teaching focused on R&P and developing understanding] is just not possible. And then if the teacher wants their pupils to be successful, if they're heading to a secondary school, then they have to do without that [R&P]. So [the teacher] just has to adjust to it.
>
> *(CZ-pre)*

Standardised assessments were also perceived as impediments by Turkish and Slovak PMTs. They emphasised that solving many tasks did not require R&P skills.

> Since the focus of pupils is always on HSE [high school entrance] exam and, you know, written exams, and in HSE exam and written exams, rather than this... proof ... pupils start to focus on 'How can I solve more questions?', 'How can I practice more?', 'How can I memorise this?'
>
> *(TR-pre)*

> It's just a question of whether there is space, whether there is time... to align it with the curriculum and everything else... The education system is such that we try to prepare pupils for graduation exams and standardised tests, but those exams don't really depend on whether pupils can justify their answers.
>
> *(SK-pre)*

This analysis shows, among others, that before the Intervention, some PMTs perceived R&P as peripheral to the curriculum rather than integral to it. Other quotes confirmed this:

> In other words, it is not like "[R&P] also needs to be described" in the objectives, but a hidden curriculum. [...] I think it is a process that is completely left to the teacher's competence. When the teacher doesn't see a pupil willing to do this or doesn't feel competent, they probably can't do it, and they probably can't even incorporate it into the lessons because of time constraints or something like that.
>
> *(TR-pre)*

> [R&P is] hardly taught at school anymore.
>
> *(IT-pre)*

At the same time, some Norwegian PMTs believed that the recently changed national curriculum supported the inclusion of R&P: "I feel it has a very big role when you ... [have reasoning] both in the subject-relevance part and core elements." (NO-pre)

(b) Time

Time was frequently mentioned as the main obstacle to incorporating R&P-related activities in the PMTs' lessons before the Intervention:

> Many teachers don't go into it [R&P] so much. Because we have the problem of catching up on a curriculum, it takes a lot of time for the pupil to question, to understand, and to do.
>
> *(TR-pre)*

Some PMTs were concerned that R&P activities would potentially decelerate the progression of the lesson, particularly when accommodating pupils with diverse skill levels, which could result in disengagement.

> If you take a very long time, then maybe more pupils will drop out, if you also have a big leap between them, like very strong and then not so strong pupils. Erm, maybe, yes, part of the class will stick with it, and then the rest will fall away, and then they will perhaps lose their desire to learn and sense of mastery, and then there will be a negative attitude towards the subject. Which can be difficult to reverse.
>
> *(NO-pre)*

Nevertheless, the PMTs articulated several justifications in the pre-interviews why it was still valuable to invest time into R&P, grounded in the conviction above that the integration of R&P activities enhances comprehension and retention of mathematical concepts. For instance, the Czech PMTs deliberated on the advantages and drawbacks of emphasising conceptual understanding – linked with R&P – over procedural learning. They generally favoured the former and largely concurred that incorporating R&P into educational practices, which enhances pupils' comprehension, ultimately saves time, as it mitigates the need for repeated re-visitation of fundamental and misconstrued concepts.

Time remained among the most frequently mentioned issues associated with teaching R&P after the Intervention, but following their experience with lesson planning and enactment, the PMTs were more specific about their concerns. First, the time required to plan an R&P lesson was mentioned, as one must consider possible pupil responses in advance. Second, the time cost of an R&P lesson was associated with the need to allow all the pupils in the class to reason: "You don't have time if you're alone. You don't have time to elicit reasoning if 70 percent of the class struggles with it" (NO-post), "reasoning might be difficult for pupils. Teachers need time, time, time" (IT-post). Third, the PMTs realised the need to build socio-mathematical norms favourable to R&P, which, again, takes time: "[R&P] has to be worked on over a long period of time." (NO-post) Fourth, some Czech and Slovak PMTs worried that teaching R&P is time-consuming and that the time costs may be extensive, as no apparent learning gains are seen immediately after the lesson.

> It is incredibly time-consuming [...] so if I had to choose only one extreme, it would rather be the hard work of memorising it, because at least a pupil can apply it, but if they only learn the principles, they can't calculate.
>
> *(SK-post)*

Unlike the rest, some Turkish PMTs did not see time as a limit any more after implementing the R&P lesson.

> I was always thinking like this, also in the method course: "You know, how are we going to do these things? We don't have this much time. How will we give the logic of everything? How will we make [pupils] discover?" and so on. It happens when you want it to. You know, I think it is really more effective, more instructive, and more beautiful when you plan it well and involve the pupils. I mean, time would be enough.
>
> *(TR-post)*

(c) Pupil difficulties

Before the Intervention, many PMTs cautioned that pupils may not understand R&P. "I actually think the level [pupils' ability] is a bit low and I think the tasks that are given [...] are too difficult for them" (N5.1-pre), or "when we are unable to provide the proof because proving it becomes too difficult for them [pupils], and then you get into [saying] that's how it is" (NO-pre).

> Really, how will using this [R&P method] so much contribute to the reasoning of these children? If this child is really behind in math, if there are things that they still can't handle fundamentally, how are we going to add this on top of that?
>
> *(TR-pre)*

As stated earlier, Italian PMTs would place R&P activities (mostly associated with proving) in secondary schools. Some Slovak PMTs suggested the same, believing R&P was more appropriate for older pupils with stronger abstract and mathematical thinking skills. There appeared to be an agreement among PMTs in the pre-interviews that R&P was primarily designed for pupils considered intellectually advanced or exceptionally talented, rather than for those identified as less academically proficient.

> Well, I guess we should take-into-account those pupils who [...] might not have the capacity to even think about why and are happy to at least know that it is so.
>
> *(CZ-pre)*

> In some schools, teachers suggest a more in-depth treatment of R&P only for the brightest pupils. […] In my class, R&P in this sense was only used in geometry for a few, the better ones, to teach them to reason and not to trust evidence. The other pupils didn't understand much.
>
> *(IT-pre)*

> [In my class], there are some smart kids who like downright seem to need [R&P], yeah. Some of them really need to know why it's like that, and if it's not like that, they almost protest that they're not going to do it until they figure it out themselves or it's explained to them somehow, so it's important.
>
> *(CZ-pre)*

However, despite admitting that R&P presents challenges for pupils and that they might be incapable of engaging actively with the proofs, some PMTs still perceived the inclusion of proofs as important, even if pupils did not carry them out independently:

> It doesn't matter – if the teacher proves it or if the pupils prove it, but the proof itself should be there. If the pupils can figure it out, so much the better; if they can't figure it out, it should at least be supplied by the teacher.
>
> *(CZ-pre)*

After the Intervention, some PMTs remained convinced that R&P was too difficult for some pupils. Czech PMTs even expressed concerns that the heterogeneity in pupils' abilities in one class would make it impossible to implement R&P in their teaching. They tried to find ways to overcome this problem (suggesting mostly group work in which more able pupils would help the weaker ones).

Yet, some PMTs changed their mind and agreed that it was not true that "proof is for the talented few" (IT-post), but rather that certain types of arguments were suitable for different levels of ability.

> Everyone has to try to achieve R&P as a teaching objective, albeit in different forms.
>
> *(IT-post)*

> … awareness that an argument can be set at different levels, and at each developmental stage, a certain type of argument is appropriate. So, for example, what to expect from an eighth-grader versus a secondary school pupil, and so on.
>
> *(SK-post)*

This change in perspective might stem from the PMTs' realisation that R&P was not deductive proof only, or from their experience with the enactment of the lesson, which they jointly prepared during the Intervention:

> I didn't know that pupils were so open to thinking... some pupils can really, uh, when we give them a good thinking time and when we give them the necessary materials, they can really do proof. They can defend their thoughts. They can explain where they came from, how they came to their conclusions. I was surprised by this, to be honest.
>
> *(TR-post)*

(d) Establishment of appropriate socio-mathematical norms in classes

Before the Intervention, the PMTs did not attribute pupils' struggle with R&P solely to pupil ability, but also to their prior school experiences and something which we would call socio-mathematical norms (Hiebert et al., 1997; Stephan, 2020) in mathematics lessons ("I think, ideally, there should be a culture in mathematics teaching." (NO-pre)). Such norms were perceived to be established in earlier school grades; however, this was often not the case. According to some PMTs, lower grade teachers would not take R&P into account: "The pupils are not prepared for this [R&P] at the elementary level." (CZ-pre) Pupils not being accustomed to working with R&P was discussed by several Norwegian PMTs also in relation to their future teaching:

> I actually experience quite a lot, at least at the lower secondary level, that it is ... they don't have that much [...] grounds to be able to say why and to be able to reason to arrive at the correct answer. [Pupils] are a little too locked into the algorithm's world.
>
> *(NO-pre)*

> [...] because you are in the eighth grade, then you receive a class that is not used to generalisation and [...] even if you have the desire to get there, then it is ... if you work on it for three years, still it is not certain that you will get there.
>
> *(NO-pre)*

Motivation for R&P was an important part of favourable socio-mathematical norms. Some PMTs reflected on its absence, also in connection with their own experiences.

> [Lack of R&P activities in schools] is due to a lack of time, but also to pupils' "lack of interest" in mathematical proof.
>
> *(IT-pre)*

And I know I was pretty much the only one who was comfortable with [R&P], that most of the class didn't like it and would rather have a parallel teacher who was more into the drill.

(CZ-pre)

I don't recall ever as a pupil myself feeling the need to prove anything, I guess it was more that I didn't mind taking things as they were.

(CZ-pre)

After the Intervention, the "missing norms"[3] were seen as problematic by many PMTs. According to them, pupils were not used to reasoning; R&P was "almost like a new concept" for them (NO-post). The Czech PMTs warned that pupils' motivation towards R&P might be low, as in the elementary years they accepted the validity of a statement at face value and were not asked to justify it. The role of the teacher was seen as important, as they needed to create opportunities for reasoning through careful questioning and interaction, rather than expecting it to happen naturally during a lesson without encouragement.

Reasoning is not something that just comes as a kind of by-product, but it's something you as a teacher have to fish for, or you have to, like, you can't just give a task and expect it to happen; you have to draw it out of them. Yes, you have to ask [pupils] good, the right questions so that reasoning and proof come through.

(NO-post)

Reasoning is not something that just appears. It is something that you, as a teacher, must practice, encourage, and incentivise. In my opinion, it's not even something you can teach, you have to make it become a need.

(IT-post)

(e) PMTs' lack of prior experience and pedagogical content knowledge

Some PMTs perceived themselves as lacking experience with R&P as learners, and they posited that this deficiency would adversely affect their capacity to carry out R&P activities:

Although I see [R&P] as important, I find it difficult at the moment to include it later [in teaching] because I haven't gone through it myself [as a learner].

(CZ-pre)

After the Intervention, some Czech PMTs worried that their knowledge (which we would call pedagogical content knowledge) might not be deep enough in terms of selecting the appropriate level of R&P for the class and reacting to pupils' suggestions and reasoning. Similarly, seeing how complex planning R&P lessons was and reading some articles about it, a Turkish PMT claimed that as her knowledge increased about the subject matter, she realised that it was more complicated than she thought, that it was "above [her] level at the moment" (TR-post), even though she was still convinced of R&P's importance for pupils' mathematical understanding.

The preceding text documents in detail how PMTs understood R&P before and after the Intervention. The changes observed in their beliefs about R&P in teaching will be summarised in Section 6.1.

5.2 Lesson plan designs

This section zooms in on the core activity within the Intervention, namely, lesson planning. In Section 5.2.1, we explore Version 3 of each country's lesson plan (LP) to determine their similarities and differences. Lesson planning was also part of the Student Meeting, during which PMTs worked in mixed international groups and jointly prepared and presented one LP in each group. These LPs were also analysed and compared, similarly to national LPs (Section 5.2.2). In Section 5.2.3, we return to the national LPs and describe how these plans changed from Version 1 to Version 3 and what might have caused the modifications.

5.2.1 Characteristics of each country's lesson plan

In this section, we first describe and interpret the content of Version 3 of the LPs of the five countries (Section 5.2.1.1) and then identify similarities and differences between the LPs (Section 5.2.1.2). The process through which Version 3 originated is depicted in Figure 4.1 in Section 4.1.2.

5.2.1.1 Description and characteristics of lesson plans

Following the analysis described in Section 4.3.2, for each country's LP (Version 3), we first (A) describe the content of the LP, and (B) provide our interpretation.

Description and characteristics of the Czech lesson plan

Lesson for the 9th grade: Simplification of algebraic expressions (A) Although the rationale and content of the Czech LP were partially swapped, the PMTs stressed in it that "justifying is crucial in mathematics" and that algebraic expressions should not be perceived by pupils "only as a cluster of letters", but that pupils should "see something deeper behind them". The lesson aimed to justify the correctness or incorrectness of proposed solutions regarding algebraic expression simplification, as well as to "visually or verbally interpret algebraic expressions". This was mediated via a worksheet with three parts and by organising a 45-minute lesson into four parts: (1) "evocation", in which pupils should have assessed proposed solutions and justifications and found mistakes – worksheet part 1 (Figure 5.1 left) (15 min); (2) "think-pair-share" (a strategy of collaborative learning), in which pupils should have identified the common mistakes and justified why the solutions were incorrect – worksheet part 2 (Figure 5.1 right) (10 min); (3) "group work", in which pupils should have linked the algebraic expressions with their graphical representation and vice versa – worksheet part 3 (see Appendix 5) (10 min); and (4) "reflection", where pupils should have shared on the blackboard the representations from part 3 and linked them with justifications used in part 2 (10 min). In the fourth part of the lesson, PMTs also planned pupils' self-assessment (using a scale) regarding fulfilment of one of the lesson's goals.

Regarding materials and resources, PMTs mentioned the worksheet and "pupils' willingness to share their ideas ... [ ] because they learn from each other". It was assumed that pupils would "learn from each other while working in pairs or a larger group. The teacher will only be the assigner of the activities and possibly help in Part 3". The teacher's role was to "assign activities precisely and clearly" and answer pupils' questions. However, PMTs did not list any possible questions/scenarios and corresponding specific teacher reactions/moves. Nevertheless, they anticipated a few ways of pupils' reasoning (e.g., appeal to authority, providing a counterexample). PMTs also outlined two subsequent lessons, building on the assigned homework tasks, focusing on the link between symbolic and graphical representations of algebraic expressions, using algebra tiles. Moreover, pupils were expected to design/choose (as part of the homework) an algebraic expression "for which some mistakes can be expected from your classmates".

(B) The Czech LP (and the subsequent outlined lessons) was designed to be used in schools where mathematics is taught according to the Hejny method[4], which (according to PMTs) means that pupils are used to working collaboratively and sharing their ideas. Hence, the LP was clearly pupil-centred, emphasising pupils' active participation, engagement, and collaborative learning. At the same time, the teacher's role as facilitator was not sufficiently articulated, and it is unclear to what extent this role was thought through (e.g., which questions to ask). In any case, no teacher moves were listed in the LP. In addition, pupils' prior knowledge and possible difficulties were addressed only briefly, without a deeper analysis. Except for the worksheet, no other resources were mentioned in the first lesson; algebra tiles were mentioned for the subsequent lesson/s. Hence, the quality and use of the worksheet were at the core of the lesson. Without knowing about pupils' prior knowledge and

PART 1 – "Are solutions and justifications OK?"

A) Simplify: $\dfrac{4a+b}{2}$

Makar's solution with justification:	Aida's solution with justification:
$\dfrac{\cancel{4}a+b}{\cancel{2}} = 2a+b$  I have checked the correctness for the pair $a=3, b=0$, next $a=0$, $b=0$ or $a=4$, $b=8$. I would certainly check it for other pairs.	$\dfrac{4a+b}{2} = \dfrac{\cancel{4}ab}{\cancel{2}} = 2ab$  I have checked the correctness for the pair $a=\frac{1}{2}, b=2$, next $a=0$, $b=0$ or $a=\frac{1}{3}$, $b=4$. So it also holds for fractions.

B) Simplify: $7a^2-3a-(1-a)$:

Achiles's solution with justification:	Izolda's solution with justification:
$7a^2-3a-(-a+1) = 4a+a-1 = 4a^2-1$ Minus before a bracket changes the signs into the opposite ones. I have simplified the powers according to the rule from the textbook. 	$7a^2-3a-(-a+1) = 7a^2-3a+a+1 =$ $= 7a^2-4a+1$ I was careful about the minus before the bracket. I have added the terms with a but I cannot simplify the expression any further. 

PART 2 – "Are the simplification correct? Why?"

a)	$\dfrac{(a+b)^2}{a^2+b^2} = \dfrac{\cancel{a^2+b^2}}{\cancel{a^2+b^2}} = 1$
b)	$\dfrac{(a-b)^2}{a+b} = \dfrac{a^2-b^2}{a+b} = \dfrac{\cancel{(a+b)}(a-b)}{\cancel{a+b}} = a-b$
c)	$\dfrac{5^2-4^2}{3} = \dfrac{(5+4)(5-4)}{3} = \dfrac{\cancel{9}\cdot 1}{\cancel{3}} = 3$
d)	$(a-b)^2 - 2\cdot\left(-ab-\dfrac{b^2}{2}\right) = a^2 - \cancel{2ab} - \cancel{b^2} + \cancel{2ab} + \cancel{b^2} = a^2$

FIGURE 5.1 The Czech LP's worksheet.

potential misconceptions, it was not easy to evaluate. However, "internally" (i.e., looking at the internal coherence of the worksheet), there were some discrepancies between part 1 and part 2. While part 1 addressed "checking with numbers" and the importance ("duality"?) of the minus sign, part 2 did not follow on from this, as many other typical misconceptions with algebraic simplification were included (e.g., algebraic identities). Nevertheless, R&P was clearly at the core, as pupils were expected to assess flawed or incomplete justifications related to simplifying algebraic expressions, to identify errors in provided solutions, and to justify them (preferably with a geometric representation), aiming at conceptual understanding. In this way, PMTs assumed pupils would, in the future, avoid the most common errors related to manipulation of algebraic expressions. Overall, in the enacted lesson, reasoning would be conceptualised mainly as using logical arguments for validating conjectures, and in terms of paths of reasoning, PMTs strove for a transition from appeal to authority or empirical reasoning to deductive reasoning/ deduction using a model. However, this transition was not explicitly articulated in the LP.

The three stated learning goals were quite ambitious given the time allocated (45 minutes), hence their time planning was unrealistic. However, PMTs probably meant the three learning goals to "cover" three lessons, and the LP was for one of the three. The other two were addressed in subsequent lessons, again by using manipulatives (algebra tiles) and by discussing new algebraic expressions designed by pupils in the homework (pupils' engagement and activation). However, it was not clear how the teacher would elicit evidence related to these two learning goals.

Description and characteristics of the Italian lesson plan

Lesson for the 10th grade: Introduction to non-Euclidean geometry (A) Concerning the rationale, the Italian PMTs suggested juxtaposing non-Euclidean with Euclidean geometry (also epistemologically) to show that different "environments" need different approaches to proving. They highlighted the importance of developing a deductive approach to proof and referred to a critical axiomatic approach to construct/understand a theory. PMTs discussed how this approach could improve pupils' problem-solving strategies in mathematics, but also in real life. In particular, they referred to geometry and how, by changing the "geometrical environment", the properties of mathematical objects change and in what way it is possible to argue/prove them ("What happens if …?" "Why?"). In line with this, PMTs mentioned, in the objectives of their LP, the role of R&P in mathematics.

The plan was a series of five lessons designed for 10th-grade pupils attending non-compulsory afternoon teaching sessions. Each lesson was

60 minutes, divided into three phases: 10 minutes for opening, 40 minutes for the middle phase (laboratory activities and open discussion), and 10 minutes for closing and final discussion in the whole group. As outlined earlier, the overall content of the lesson series focused on non-Euclidean geometry, particularly hyperbolic models (Beltrami-Klein and Riemann-Poincaré models), along with their key mathematical concepts and theorem proofs.

The listed lesson series objectives are related to the Italian National Curriculum and the UMI document *Matematica 2003* (Anichini et al., 2004), both in terms of R&P and the mathematical content in the thematic nucleus "Space and Shapes". PMTs even referred to these documents explicitly. Regarding materials and resources, they planned the use of GeoGebra as a "tool for the discovery of Beltrami-Klein's model", both during the teacher's instruction (Figure 5.2) and in pupils' explorations. They also incorporated the history of mathematics as an "introduction" to the topic in the form of a PowerPoint presentation. Additionally, they planned to provide notes on the Beltrami-Klein model, GeoGebra macros, and mathematical vocabulary to support and enhance pupils' mathematical language.

In the parts when "new mathematics" was introduced to pupils, the teacher presented the content (in a lecture style), while during the discussions, the teacher's role was that of facilitator. Regarding assessment, the PMTs distinguished between short-term and long-term evaluations, focusing primarily on the mathematical content rather than the didactical aspects. However, they did not propose specific methods for assessing pupils' understanding. Nevertheless, before the first lesson, a questionnaire was planned to assess pupils' geometric knowledge of the Euclidean model.

(B) This LP was very ambitious regarding mathematical content. It appears very detailed concerning the declared aims and objectives (referring to aims, knowledge, competencies, etc., in the Italian mathematics curriculum). In the five designed lessons, PMTs highlighted the importance of "point/s of view",

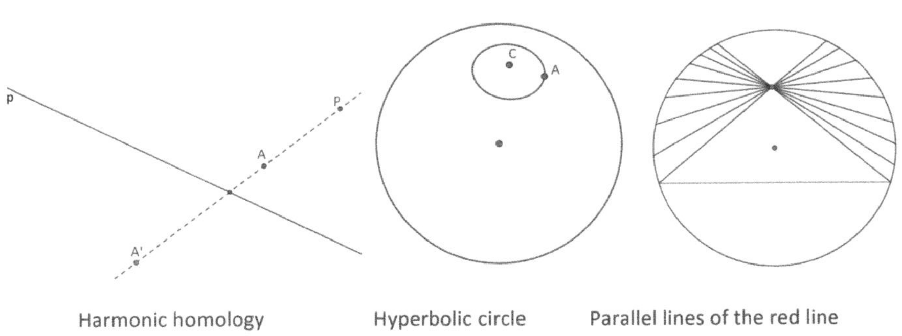

FIGURE 5.2 Constructions in GeoGebra used in the Italian LP related to the Beltrami-Klein model.

discovering how/why the properties of some mathematical objects might change according to the considered "geometrical environment", and in which way it would be possible to prove them. The PMTs stressed these aspects, discussing the passage between Euclidean and non-Euclidean geometry. They referred to the Beltrami-Klein model and the R&P evidence in this model. In the enacted lessons, reasoning was primarily conceptualised as forming conjectures and validating them using logical arguments, particularly in the closures of the lessons. During the launch phase, reasoning was associated with solving problems and communicating thinking. Given the complexity of the mathematics subject and the school grade in which these lessons were enacted, Italian PMTs often conceptualised reasoning in their LPs as the process of connecting aspects of mathematics through logical deduction. This involved reasoning that logically followed from previous assumptions, hypotheses, axioms, or established knowledge. Their planned activities aligned in this sense with mathematical definitions of reasoning concerning logical argument and deductive thinking.

The structure of the plan was coherent with the typical structure of an Italian *unità didattica di apprendimento* or didactical learning unit (as stipulated in the National Curriculum, to describe teaching a series of seven lessons), designed for 10th-grade pupils attending non-compulsory lessons. In particular, it was planned and enacted in a secondary school programme called *Liceo Matematico*, which is more oriented to a higher mathematics level than common secondary schools teach. Some didactical aspects related to PMTs' planning were not explicitly explained in the LP, possibly because of the lack of experience with lesson planning and pedagogical/didactic knowledge acquired during their study programme at the university when they planned the lesson. Regarding the declared (by the PMTs) teacher approaches, these appeared to be "teacher-centred", particularly in the presentation of high-level mathematics content. In their LP, PMTs referred to selected questions for pupils and expected answers. However, they did not always explicitly outline teacher moves (Ellis et al., 2019) and norms. Interestingly, the inclusion of the history of mathematics as a resource for teaching (as an "introduction" to the topic) and the definition of the mathematical vocabulary to improve pupils' mathematical language were addressed. These particular choices seemed linked to the approaches in some Italian secondary-level textbooks.

Description and characteristics of the Norwegian lesson plan

Lesson for the 8th grade: Figural patterns (A) In the rationale for this lesson series, the PMTs explicitly mentioned the Norwegian National Curriculum and stated that "pupils should develop their ability to see the big picture", which would "lay the foundation for further work on seeing connections, drawing logical conclusions, and reasoning about own and other people's solutions". The latter directly reflects the Norwegian National Curriculum and its core element Reasoning and Argumentation. The lesson aim was for pupils to describe and generalise figural patterns and to "reason about one's own and others' strategies".

The content of this lesson was in the field of algebra, namely figural patterns (represented by "dotted" figures, see Figure 5.3). Building on pupils' prior knowledge of simple figural patterns and basic algebraic concepts, in this LP the PMTs designed a task that required pupils to predict subsequent figures in the pattern, determine the number of circles in specific figures, use multiple methods to find solutions, create a general formula to represent the pattern, justify why the formula works for all figures in the sequence, apply the formula to calculate the number of circles in given figures, and determine the truth value of a mathematical statement related to the number of circles in a specific figure. The pupils mainly worked in groups of two. The teacher's task was to probe for understanding of what pupils had done and ask them to explain their methods. The LP also involved anticipated pupils' misconceptions, strategies, and difficulties. In terms of assessment, both informal (e.g., a teacher observes pupils) and formative assessment were outlined (e.g., an oral review of pupils' strategies in plenary, exit cards). Regarding resources, the following were mentioned: blackboard, stationery, calculator, notebook, and curriculum materials, such as previously prepared task sheets and homework sheets. The lesson was 45 minutes long, divided into three phases: start-up, main part, and closure, with approximate timings attached.

In compliance with the aim related to the description of figural patterns (by pupils) in their own words, R&P was addressed when pupils had to generalise figural patterns, see connections between different figures, and reason about their own and others' strategies for generalisation.

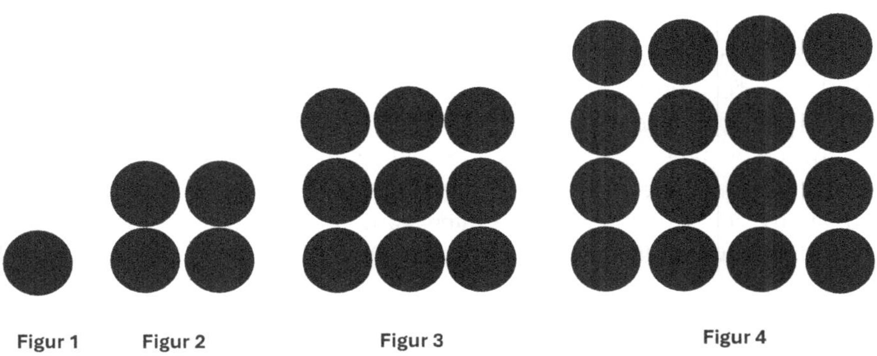

FIGURE 5.3 Figural pattern used in the task in the Norwegian LP.

Teacher moves were identified using Ellis et al.'s research (2019), including the facilitation of dialogue for pupil learning by the teacher, for example, "by encouraging them [pupils] to correct mistakes and in the plenary discussion represent pupils' thoughts", or by pupils sharing different ways and solutions and explaining "why the strategy works and possibly why it does not work".

(B) What characterised this LP (including the subsequent outlined lessons) was the emphasis on communication (e.g., sharing different strategies), both between pupils (in groups of two) as well as among pupils and the teacher (e.g., plenary; teacher walking round and helping pupils). There was a clear avoidance of difficult language for pupils, for equity reasons, so that everybody should be able to do something during that lesson, regardless of pupils' algebraic background knowledge, and a purposeful use of simple language in terms of pupil learning goals. Equity reasons were also visible in the assessment part of the LP: the exit cards asked for feelings of well-being. The lesson was pupil-centred, with the aim of helping pupils become familiar with describing and reasoning about figural patterns. From the LP, it was evident that the Norwegian PMTs perceived reasoning mainly as communicating thinking, which was followed by validating thinking. In the LP, there were also traces of reasoning as forming conjectures and reasoning as using logical arguments to validate conjectures. The latter mainly concerned the selected task, since in terms of paths of reasoning, one could expect inductive reasoning (used when inferring a general rule), as well as simple one-step deduction (used when justifying the truth value of the given statement). Overall, it seemed that the lesson was more inspired by pedagogic moves rather than mathematics didactical considerations.

Description and characteristics of the Slovak lesson plan

Lesson for the 9th grade: Application tasks leading to linear equations (A) The Slovak PMTs highlighted in the rationale of their LP the importance of linear equations and inequalities, as they considered them "basic building blocks in all fields of mathematics, as well as applications of mathematics in practice". For this, a "certain level of abstraction" was needed, and the main activity was expected to lead pupils to it.

The lesson aimed at solving simple application tasks (number pyramid, see Figure 5.4) with specific rules, and "without formalisations into the form of equations", choosing a suitable strategy, and at verification of the solution's correctness, while the latter two aims were related to the Slovak National Curriculum. However, in the content section of the LP, PMTs also explicitly mentioned the justification/reasoning of individual steps in the solution process, as well as "solving tasks that require procedures that are different from the usual procedures/algorithms".

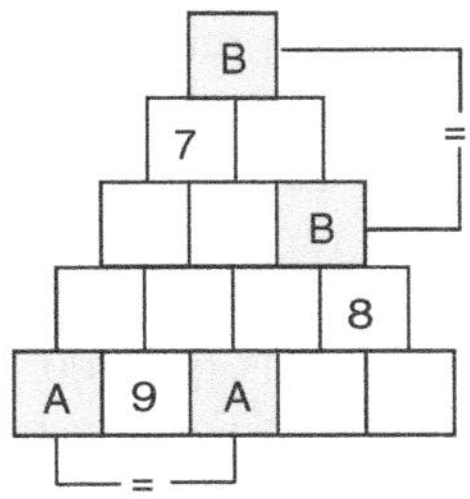

FIGURE 5.4 Number pyramid used in the task in the Slovak LP.

Regarding materials and resources, a printed pyramid and number cards that fitted in the boxes of the pyramid, a whiteboard with markers, notebooks, and pens/pencils were expected to be used.

PMTs planned for the whole activity two 45-minute long consecutive lessons and divided it into several parts: introduction (5 minutes for greeting and bureaucracy and 10 minutes for the introduction of the task and motivation/link to the real-life situation supporting its meaningfulness), main part – "problem-solving", composed of pupils' individual work (15 min), work in groups of three (20 min), whole classroom discussion led by the teacher (30 min), and "summarisation" by the teacher (7 minutes to summarise main ideas and 3 minutes to assign homework). During the individual work, pupils were expected to find "a way (step by step) to solve the task" and later in the group work, to share their ideas and to discuss the correctness and effectiveness of the proposed solutions. In the whole-class discussion, where the teacher was in the role of a guide, the focus was on "thoroughly justifying the individual steps of the solution", for which PMTs explicitly listed several teacher moves (Ellis et al., 2019) to support pupils' reasoning. Regarding assessment, PMTs planned to provide formative feedback to pupils.

(B) To characterise this LP, there is a mismatch between the lesson as it was conceived/designed and PMTs' intended aim (i.e., to introduce linear equations). In other words, the LP was about pre-algebraic thinking (e.g., notion of variable, unknown) that, according to the Slovak curriculum, has been introduced in lower grades (grade 8). Moreover, in this lesson, there was no explicit link to linear equations (which are introduced in grade 9). At the same time, PMTs planned to have the teacher formally define the concept of a linear equation and demonstrate it through examples in the upcoming lessons. In the following, we characterise the LP as it was presented.

The nature of the pyramid task appeared to encourage exploration and the gradual discovery of the main idea for solving the task, which requires analytical thinking and deductive reasoning. PMTs facilitated this through various forms of work (individual, group, and whole class) and provided a sufficiently long time for pupils to work on the task (2 × 45 min). The LP showed that the PMTs' perception of reasoning primarily involves conceptualising reasoning as problem-solving, followed by reasoning as validating thinking. While all other parts of the provided spider web template were sufficiently elaborated upon, the method for pupil grouping was missing. Regarding assessment, it was not articulated how pupil learning could be assessed. A justification for materials (printed pyramid with number cards) was also missing, although PMTs very likely made this choice consciously. Expected paths of reasoning were implicitly articulated in the sections "pupils' role" and "teacher's role", moving from empirical reasoning to deductive reasoning.

Overall, the Slovak LP aims for advanced deductive reasoning through a task that could be motivational (but also demanding) for 9th graders, it allocates sufficient time and working forms, and suggests several Ellis et al.'s (2019) teacher moves (e.g., eliciting ideas, pressing for explanation).

Description and characteristics of the Turkish lesson plan

Lesson for the 8th grade: Algebraic identities (A) Concerning the rationale, the Turkish PMTs emphasised in their LP the importance of not only memorising the formula of an algebraic identity, but also understanding its logic, for example by relating it to a different representation or expressing it in different semiotic registers. This was said to also help pupils with later mathematics curriculum topics (e.g., factorisation).

The lesson's aim was connected to the Turkish National Curriculum's 8th-grade objective of "explaining identities with models". Specifically, the aim was to link the content, the algebraic identity $(a+b)^2 = a^2 + 2ab + b^2$, to different representations, also using physical algebra tiles and the virtual ones in Mathigon[5] (see Figure 5.5). This was done to forestall misconceptions (such as

FIGURE 5.5 Expected pupils' solutions using algebra tiles and Mathigon in the Turkish LP.

$(a+b)^2 = a^2 + b^2$), and link it to real-life problems (e.g., a square-shaped garden photographed by a drone). In this part, we found the only reference to R&P: "Since they will prove this equation themselves, their abilities, such as reasoning and making inference, can develop." The LP also mentions experimentation: "Children who obtain the square shape learn the algebraic expression $(a+b)^2$ through experimentation."

Regarding materials and resources, the teacher-prepared activity sheet, Mathigon, algebra tiles, pencil and paper, the smartboard, and the exit cards were used.

The lesson was 40 minutes, divided into three phases: 10 minutes opening, 20 minutes "middle", and 10 minutes closing. In class, pupils were expected to work mainly in groups of three on their own strategies. The teacher was perceived as a facilitator of the activities and a guide of the plenary at the closing of the lesson. Regarding teacher moves, PMTs explicitly outlined many teacher moves (e.g., questions to be asked, including expected pupils' answers, group division, extra tasks, and resolutions about misconceptions) and norms. In terms of assessment, the PMTs distinguished between (a) "short-term" and (b) homework/longer-term assessment. For (a), PMTs designed an exit card (distributed at the end of the lesson), where pupils could freely identify and write down their difficulties in words, whereas for (b), they designed a homework sheet with exercises, where pupils had to represent algebraic identities with the geometric representation model.

(B) This LP was very detailed and meticulously thought through in every respect: clear aims and objectives (referring to the objectives in the Turkish mathematics curriculum); questions the teacher would ask and expected answers to the questions; prior knowledge and potential misconceptions; how the teacher selected the groups; what materials and resources are needed and why, etc. Moreover, the lesson reflected a belief in the beneficial use of different representations and semiotic registers, so that pupils could understand "by seeing, touching and listening" (with, e.g., algebraic tiles, Mathigon, formulas). Furthermore, the Turkish PMTs expressed their confidence that pupils could develop other identities similarly, with one algebraic identity being treated in a detailed way. This was expected to help pupils not rely on memory, but to make connections. Thus, in the enacted lesson, reasoning would be conceptualised mainly as connecting aspects of mathematics and validating thinking, evident via examples of expected dialogues in the class and the teacher's moves. Regarding paths of reasoning, the PMTs assumed pupils' empirical reasoning followed by generalisation, and they also provided an example of proof by counterexample. However, they mainly gave examples of pupils' deductive reasoning when justifying the algebraic identity through visual and symbolic representation, but they did not explicitly name the type of reasoning involved.

5.2.2 Similarities and differences across the countries' national lesson plans

Notably, all but one LP "only" implicitly addressed R&P, except for the Italian LP, which worked at a higher level (grade 10 high achievers) than the other four LPs. In many ways, the Italian LP differed from the others (e.g., length of learning trajectory considered – seven lessons; grade level – grade 10; level of achievers considered – out-of-class lesson for high achievers). However, the difference was particularly significant at the epistemological level, where non-Euclidean and Euclidean geometry were juxtaposed to show that different "environments" needed different approaches to proving. For this, geometry was purposefully chosen, whereas the other countries' LPs used algebra as a field of study. Another important difference between the Italian and the other LPs was that the Italian LPs were mainly teacher-centred, whereas the other countries' LPs were mainly pupil-centred. This could be understood due to the Italian PMTs' inexperience with lesson planning and enactment; this was their first "try" at designing and implementing lessons, but also by their stronger emphasis on advanced mathematical content in their education. Italian PMTs completed significantly more credits in mathematics courses than their peers in other countries (see Table 3.1), which may explain why their LPs reflected higher-level mathematical content.

Although lesson planning is perceived by many (Cevikbas et al., 2024) as planning the teaching and learning, including assessment of the learning, it was noteworthy that there were few mentions and details about assessment in most of the LPs, except for the Turkish LP, where a distinction was made between "short-term" (exit cards where pupils could describe their difficulties) and homework/longer-term (homework: linking algebraic identities with geometric representations) assessment. Altogether, the LP of the Turkish group was very clear and detailed in terms of aims and objectives, misconceptions, and teacher moves (Ellis et al., 2019). The Turkish LP also included many more resources than the others; these encompassed manipulatives (algebraic tiles) and digital resources (Mathigon). Also, the Italian LP emphasised using GeoGebra for reasoning in geometry. A possible explanation for the clarity and detail of the Turkish LP is that PMTs in Türkiye begin focusing on lesson planning from their second year and regularly receive feedback from instructors (e.g., the consideration of including digital and non-digital resources in lessons).

In terms of similarities, all LPs included group work and whole-class discussions. However, the ways these were implemented varied: in the Czech, Norwegian, Slovak, and Turkish cases, pupils worked on prepared worksheets followed by whole-class discussions, often led by the teacher. In contrast, the Italian case was distinct in its emphasis on dialogic teaching (led by the teacher) throughout the lesson. It was also noteworthy that there was hardly any link to a real-life context in all the LPs.

Regarding particularities, the Norwegian LP drew heavily on communication, amongst pupils and with the teacher, and appeared more pedagogically considered rather than mathematics didactically. Pupils' well-being, inclusion, and equity considerations seemed important to the Norwegian PMTs. This orientation may be explained by the structure of Norwegian teacher education, where pedagogy and inclusive practices are strongly emphasised, and lesson planning is mainly addressed within general pedagogy rather than mathematics education courses. A particularity of the Czech LP was the expressed mention of the Hejny method, where the teacher's role is that of a facilitator throughout the lesson. However, it was not clear from the Czech LP what the teacher (particularly a newly qualified teacher) would have to consider in class (besides preparing the worksheet). It could be that the Hejny method is clear on that, and that PMTs did not see the need to write this down. In the Slovak LP, there was a mismatch between the stated aims of the LP (i.e., introduction to linear equations) and the task provided (supporting pre-algebraic thinking: notion of variable, unknown). However, linear equations were expected to be taught and learned in subsequent lessons. Albeit not explicitly stated, the task provided in the LP asked for analytical thinking and deductive reasoning. Slovak PMTs, like the Italian PMTs, lacked experience designing and implementing lessons, which might explain inconsistencies in their LP.

Zooming in on R&P, across these LPs, R&P was addressed to varying degrees and through different conceptualisations. While Slovak PMTs conceptualised reasoning as problem-solving and validating thinking, the latter was also evident in the Turkish and Norwegian LPs. However, in the Norwegian LP, the emphasis was on communicating thinking, whereas in the Turkish LP, it was on connecting aspects of mathematics. In the Czech LP, reasoning was primarily perceived to be using logical arguments for validating conjectures. Given the complexity of the topic in the Italian LP, reasoning was associated with connecting aspects of mathematics through logical deduction. Regarding paths of reasoning in other countries' LPs, in Slovak, Czech, and Turkish LPs, PMTs provided examples/assumed pupils' empirical reasoning, through the teacher's facilitation, expected to be moved towards deductive reasoning. In the Norwegian LP, inductive reasoning was at the core, due to the nature of the learning activity, the generalisation of a figural pattern.

5.2.3 Internationally mixed groups' lesson plans

LPs developed during the Student Meeting will be presented in the same way as the national LPs described earlier. The composition of the groups of PMTs working on the same LP is given in Section 4.2.2.

5.2.3.1 Description and characteristic of the mixed groups' lesson plans

Description and characteristics of the Group 1 lesson plan

Lesson for the 8th grade: Identification, construction and classification of triangles Group 1 rationalised their 90-minute LP assuming that triangles, specifically their identification, construction, and classification, were part of the curricula across the (MaTeK) participating countries. They emphasised that triangles are "one of the main building blocks" in mathematics and that "geometry is everywhere in real life". The central aims of the lesson were to support the identification of triangles in real-life surroundings and their construction and classification, including the discovery of triangle inequality.

A distinctive feature of this lesson was the active, hands-on construction and recognition of triangles using a rope divided into parts, with learning taking place partly outside the classroom. Pupils were encouraged to work in small groups, engaging in collaborative tasks that required them to respect one another's opinions. The provided worksheet structured the entire lesson: it included tasks such as finding triangles in the environment and constructing them using the rope, which served as a tool for experimental demonstration and exploration of geometrical properties. While the paths of reasoning were not explicitly mentioned, the lesson implicitly supported inductive reasoning and discovery learning through manipulation and experimentation.

The role of the teacher was primarily to facilitate group work, stimulate reasoning, and assess pupils' thinking, particularly by probing for explanations, clarifications, and justifications (e.g., "why is this a triangle?") and by comparing multiple solutions. Formative assessment strategies were planned, including using exit cards to encourage pupils' reflection and self-assessment. However, as the final summary and discussion were also intended to take place outdoors, concerns were raised about whether the location was suitable for comparing strategies, collecting justifications, and completing exit tickets effectively. Overall, the lesson was pupil-centred and rich in opportunities for hands-on involvement, yet how pupils' developing thinking would be effectively captured and assessed – either by themselves or the teacher – remained somewhat underdefined.

Description and characteristics of the Group 2 lesson plan

Lesson for the 8th grade (13-14-year-old pupils): Area of a triangle with the support of the parallelogram This 60-minute lesson focused on guiding pupils to discover the formula for the area of a triangle by relating it to the properties of a parallelogram. Although the PMTs did not explicitly state their rationale, the underlying intent appeared to be using geometric manipulation (cutting and rearranging shapes) to support conceptual understanding. The lesson content centred on the relationship between triangles and parallelograms, including their shared properties and the sum of angles in a triangle.

A variety of creative, pupil-centred activities were planned. Pupils were asked to tessellate a parallelogram to determine its height, cut it into triangular parts, design their own parallelograms and triangles, and explore the area and angle sum within these figures. Materials provided included paper, scissors, and pens to facilitate hands-on learning and experimental demonstration. These tasks were meant to lead pupils towards forming and validating their own conjectures through logical reasoning, although paths of reasoning were not explicitly articulated in the LP.

Despite the engaging and exploratory nature of the lesson, several aspects remained unclear or underdeveloped. The teacher's role was minimal, primarily expected to probe for justification (e.g., regarding the angle sum in a triangle), raising concerns about whether pupils would be effectively guided or supported in identifying and correcting misconceptions. No formative assessment strategies were described for either the teacher or the pupils during the activities. While exit cards were suggested at the end of the lesson, they were general in nature (e.g., "How did you feel about your learning today?") and lacked focus on specific learning goals.

Although the PMTs presented their view of reasoning as involving validation through logical arguments, how pupils would be helped to reach the area formula for a triangle from a parallelogram was not clearly shown. In contrast, the angle-sum activity was better structured. While the rationale hinted at "long-term learning", how this would be achieved, especially in connection with the lesson objectives, was poorly elaborated. While the lesson ideas showed promise and creativity, more detail and teacher guidance would be necessary for a didactically robust and coherent LP.

Description and characteristics of the Group 3 lesson plan

Lesson for the 8th grade: Area of a circle Group 3's LP aimed to foster pupils' empirical thinking and informal proving, with a long-term goal of supporting future understanding of the volume of cylinders. The core objective was for pupils to discover the connection between the area of a rectangle and that of a circle, ultimately leading them to derive the formula for the area of a circle. To achieve this, a hands-on activity was designed where pupils cut a circle into four equal sectors and attempted to rearrange these into a familiar geometric shape, ideally, a rectangle. If needed, they would further cut each sector into two and continue rearranging until the resulting shape approximated a rectangle sufficiently for them to calculate its area.

This process was intended to help pupils informally derive the formula for the area of a circle, using known formulas for the area of a rectangle and the circumference of a circle, leading to the expression $\pi r \cdot r = \pi r^2$. While the teacher suggested the cutting method, pupils were expected to discover the correct shape independently and justify why the area remained constant throughout the transformation. Pupils worked in small groups of two or three, engaging in active discussion, sharing reasoning, and explaining both their constructions and conclusions.

The teacher's role was that of a guide, posing "why" questions and offering scaffolding where needed. Materials included paper, scissors, a compass, a rubber, and a pencil. Assessment was to take place through self-assessment in the final five minutes of this 45-minute lesson, along with a brief summary. However, specific assessment criteria, the content of the summary, and the division of time among the three lesson phases were not elaborated, likely due to time constraints during the planning process.

The lesson was pupil-centred and designed to promote empirical reasoning through active manipulation and exploration. Although PMTs did not explicitly identify paths of reasoning, the activity reflected an experimental approach to informal proof, aligning with the empirical proof schemes described by Harel and Sowder (2007). Anticipated difficulties and corresponding teacher responses were only minimally addressed, suggesting a need for more detailed preparation. Nonetheless, the lesson provided a creative and conceptually rich context for pupils to construct mathematical understanding through guided discovery.

Description and characteristics of the Group 4 lesson plan

Lesson for the 12-14-year-old pupils: Area of a triangle Group 4's 45-minute lesson was designed to develop pupils' reasoning skills in the context of calculating the area of triangles, with a secondary rationale related to real-life applications such as estimating surface area for painting. The stated aim was for pupils to calculate the area of a triangle in different ways and to determine whether different triangles have the same area – an idea linked to the underlying concept of triangle equivalence or identity. Although this concept was listed as a learning goal, it was not explicitly addressed in the LP.

The tasks themselves were thoughtfully sequenced. The first task involved calculating the area of four triangles placed on a grid. The second and third tasks required pupils to determine the location of a missing vertex to obtain a triangle of a given area, with variation in the orientation of the given side – horizontal in task two and diagonal in task three. Materials listed included the worksheet and grid paper (Figure 5.6).

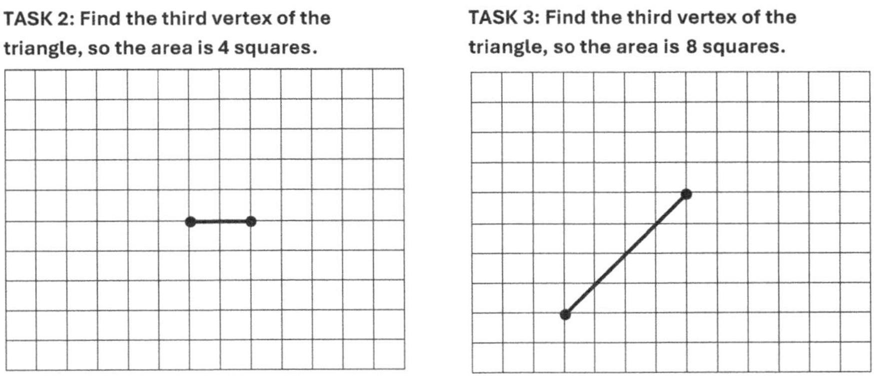

FIGURE 5.6 Designed tasks 2 and 3 in the worksheet prepared by Group 4.

The lesson structure was well organised and consistent across three interconnected, tiered tasks. Each task followed a pattern: individual work (2 minutes), group work in fours (5 minutes), followed by a whole-class discussion (5–6 minutes). The beginning of the lesson (2–3 minutes) was devoted to setting the stage, and the final 5 minutes were allocated for self-assessment via an exit card. The card included three closed questions and one open question focused on pupils' perceived challenges, but did not yield insight into pupils' actual mathematical learning.

While the tasks had strong potential to support generalisations about triangle area and encourage links between algebraic and geometric representations, these connections were not explicitly articulated among the learning goals. PMTs included relevant prior knowledge and anticipated five solution strategies for each task. They also proposed teacher questions and moves for each task to guide the whole-class discussions and to encourage pupils' reasoning, especially during the sharing phase. Since the tasks themselves did not prompt argumentation directly, the teacher's role in pressing for explanations and justifications was crucial.

Despite the lesson's rationale for developing reasoning, explicit reference to expected modes or paths of reasoning was absent from the LP. Nonetheless, the design offered opportunities for pupils to transition from empirical reasoning or reliance on authority – some of which were reflected in the anticipated strategies – towards more deductive reasoning or reasoning with a generic example. These potential developments were explored further during the PMTs' plenary presentation, where they also conceptualised reasoning as the validation of thinking through logical argumentation.

Overall, the LP was well structured and thoughtfully designed. However, it would have benefited from further elaboration of the reasoning processes it intended to develop, greater clarity around how key concepts (such as triangle equivalence) would be addressed, and more targeted formative assessment strategies. These refinements could have strengthened the plan's didactical depth and alignment with its stated learning goals.

Description and characteristics of the Group 5 lesson plan

Lesson for the 14-year-old pupils: (Approximation of the) area of a circle Group 5's 45-minute lesson aimed to help pupils "see the connection between polygons and a circle" and explore how a circle's area can be approximated without directly using a formula. PMTs rationalised their design by referencing curricular requirements and the lesson's role as a "stepping stone for other geometry". The intended focus was on discovery and conceptual understanding rather than formula application, with the aim of fostering pupils' reasoning and helping them see that "math is not random [...] there is a reason for everything, and math is connected".

The lesson followed a hands-on, pupil-centred approach. Pupils worked in groups of three with wooden sticks of equal length and paper circles. In the first task, they constructed regular polygons inscribed within a circle, and then explored how the type of polygon changed depending on the stick length. In the second task, GeoGebra was introduced to support further exploration. In the third task, pupils were expected to explain how the area of the constructed polygons could be calculated and to discuss, first in groups and later as a class, how such knowledge might help approximate the area of the circle.

The role of the teacher was that of a facilitator, supporting reasoning and guiding whole-class discussions by writing pupils' ideas on the blackboard, offering formative feedback, and pressing for justifications. The integration of a YouTube video visualising the circle approximation process further supported the shift from empirical observations to more formal reasoning.

Although the group admitted that their LP was more of a skeleton due to time constraints, the sequence of tasks was coherent with the stated learning goals. The lesson offered potential for rich mathematical thinking, particularly if supplemented by well-anticipated scaffolding and carefully timed teacher interventions. In terms of reasoning, the PMTs envisioned a progression from empirical reasoning – emerging from hands-on manipulation and visual observation – towards deductive reasoning as pupils grappled with justifying how a polygon can approximate a circle.

Their presentation showed that the PMTs conceptualised reasoning as the ability to connect mathematical ideas and use logical arguments to justify relationships. In this context, the teacher's role was crucial – not only in maintaining the mathematical focus of the discussion but also in supporting pupils to move beyond surface-level observations and towards deeper, more connected understanding. With further elaboration, particularly regarding assessment and scaffolding, this LP had the potential to place reasoning at its core.

Description and characteristics of the Group 6 lesson plan

Lesson for the 8th and 9th grades: Discovering the area of a trapezoid Group 6's LP focused on guiding pupils towards discovering the formula for calculating the area of a trapezoid. The rationale was rooted in fostering reasoning and argumentation, particularly by "breaking down the problem into smaller, simpler parts". The learning trajectory began with a short (5-minute) task aimed at recalling prior knowledge about the area of triangles, followed by two main tasks that led pupils through different methods of decomposing a trapezoid, culminating in applications of the area formula in real-world contexts (Figure 5.7).

The lesson was designed for group work, with pupils working in pairs or trios. This set-up was intended to support peer discussion and collaborative reasoning. The teacher's role was described as that of a facilitator or guide, supporting pupils' thinking during the group work and summarising key insights after each task during whole-class discussions. As a follow-up, pupils were assigned a homework task: to apply the methods used in the lesson to determine the area of a regular hexagon, using the decomposition strategies discussed in class.

The sequence of tasks was coherent with the stated learning goals and provided a clear pathway from concrete examples to generalisation. Although the PMTs did not include a specific suggestion for self-assessment, their presentation of the LP included clearly described teacher moves that aimed to

- There is a park in the middle of Prague, which has a shape of a right angle triangle of lenghts 7, 24 and 25 meters. There are 12 tulips per 1m^2 in this park. How many tulips are there in total?
    1. What methods did you use?
    2. Why did you use your method?
- Giuseppe has a field of shape as in figure 2. How much money does he have to pay per month, if the taxation is 10€/m^2 monthly?
    1. What methods did you use?
    2. Is there another way of approaching this problem?
- Giuseppe bought adjacent piece of land, so his field now looks like as in figure 3. What is the taxation now?
    1. How did you change your approach to this task compared to the previous one?
    2. Can we generalize this idea into common trapezoid?

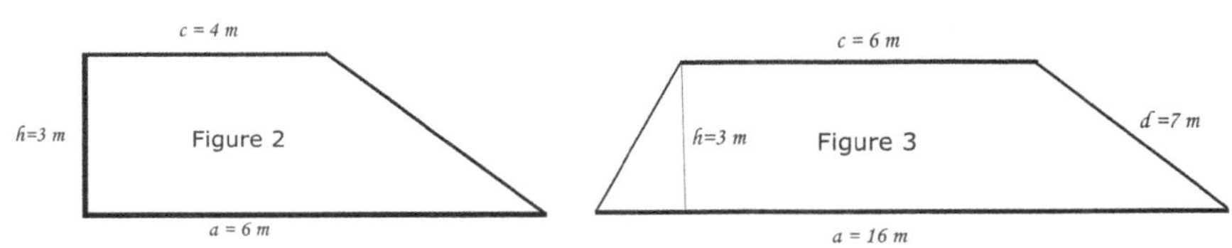

FIGURE 5.7 Designed tasks in the LP of Group 6.

support and steer pupils' reasoning. While the paths of reasoning were not made explicit in the LP, the learning experience implicitly followed a progression from deductive reasoning in specific cases to generalisation, ultimately deriving the formula for the trapezoid's area.

The group recognised this transition from specific to general reasoning as central to the lesson, even if not fully articulated in the plan itself. Furthermore, the cross-country context of implementation was noted: in all participating countries except Italy, pupils at age 14 had not yet formally encountered the trapezoid area formula. Italian PMTs, therefore, saw the potential of adapting the LP to emphasise not only the decomposition strategies but also the *justification, demonstration, and proof* of the formula's validity.

Overall, Group 6's LP was a pupil-centred design with strong potential to support the development of reasoning through guided exploration and structured generalisation, especially with further elaboration of assessment strategies and explicit articulation of reasoning paths.

Description and characteristics of the Group 7 lesson plan

6th and 7th grades: Discovering the trapezoid and its area Group 7's LP centred on pupils' exploration and discovery of the formula for calculating the area of a trapezoid through decomposition strategies. The rationale for the lesson was twofold: to engage pupils in identifying and dividing a trapezoid into simpler, familiar shapes (such as triangles, squares, and parallelograms), and to support the generalisation of the area formula through this process. The learning goal was to foster reasoning by progressing from numerical examples to an algebraic generalisation of the trapezoid area formula.

The lesson was structured into three parts within 45 minutes: a 5-minute introduction, a 25-minute group problem-solving activity based on a single rich task (Figure 5.8), and a 15-minute whole-class discussion and closure. The task context was set by a real-life scenario introduced by the teacher – helping a friend clean moss off a roof – intended to increase relevance and engagement.

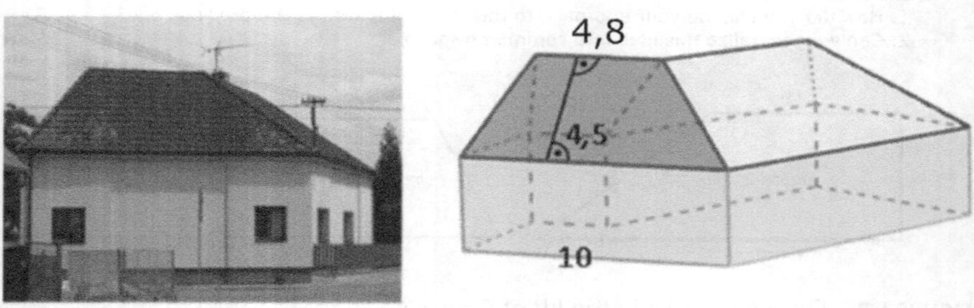

FIGURE 5.8 A task designed for a group activity in the LP of Group 7.

> Part of the roof of the family house is overgrown with moss. Under the moss, moisture is kept for a long time and the roof covering loses its quality. Therefore, the owner of the house purchased a solution used to remove mosses. The manufacturer indicated the consumption of 1/3 litre of solution per 1 m^2. The price per litre of solution is CZK 53. The dimensions of the damaged part of the roof in metres are shown in the picture. Are we able to calculate the price of the solution needed to clean the given part of the roof from the entered data?

Pupils worked in groups of two to four to foster collaborative problem-solving and discussion. The teacher's role was to stimulate interest at the beginning, guide discussions during group work, and facilitate a whole-class discussion where pupils would compare solutions and move from numerical strategies to an algebraic expression of the trapezoid area. The PMTs anticipated five different decomposition strategies, all aiming to support pupils in visualising and understanding the underlying structure of the trapezoid.

The lesson was pupil-centred and coherently aligned with its learning goals. It integrated real-world context with mathematical exploration and encouraged transitions from concrete examples to generalised reasoning. Notably, Group 7 was the only group to explicitly articulate multiple anticipated reasoning modes in their LP, starting with experimental demonstration and progressing to deduction using both specific and general cases.

However, despite this promising framework, several elements were underdeveloped. The teacher's specific moves were not well elaborated in the LP or its presentation, nor were strategies for assessing pupils' learning or addressing possible misconceptions. These limitations reduced the clarity of how reasoning would be scaffolded in practice. Nonetheless, both the LP and its presentation demonstrated that the PMTs conceptualised reasoning as the process of validating thinking through logical argumentation and the justification of conjectures.

With more time to develop the instructional details – particularly teacher questioning, scaffolding strategies, and assessment – the LP had strong potential to support pupils in constructing a grounded understanding of the trapezoid area formula through meaningful mathematical reasoning.

5.2.3.2 Similarities and differences across the mixed groups' lesson plans

All LPs emphasised the importance of pupil-centredness for learning, where pupils could actively explore, discover, and discuss their findings. In four LPs, pupils engaged with the content through hands-on activities; in other words, experimental demonstration would be a typical mode of reasoning in these lessons, although this was not explicitly mentioned in any of the LPs. In most

LPs, the teacher was positioned as a facilitator who guided pupils through the learning process, asking probing questions, encouraging deeper understanding, and pressing for justification. This guidance was to help pupils move towards deductive reasoning or to navigate pupils through informal proof related to the experimental demonstrations. The desired path of reasoning was explicitly mentioned only in one of the LPs. Still, from the PMTs' presentations in the plenary session, it was evident that reasoning was implicitly incorporated in worksheets, tasks, or in teacher facilitation.

Group work was common across all LPs, with pupils often working in small groups to facilitate discussion, peer learning, and collaborative problem-solving. In all LPs, pupils' misconceptions and difficulties were hardly considered (possibly due to the lack of time given to preparing LPs). Digital tools were not included; only one group referred to GeoGebra and YouTube for a subsequent lesson. Group 1 was unique in planning the main learning activity outside the classroom, whereas the other LPs were set in the classroom. In terms of assessment, those groups who managed to address it within the limited time all planned self-assessment through exit cards. In the LPs of Groups 1, 4, and 7, PMTs used real-world context as a starting point for "motivational" purposes. Groups 6 and 7 also designed tasks with a real-life context, providing opportunities for pupils to engage and connect mathematics with real life.

Regarding content, the instructions given to PMTs were to design a lesson related to geometry and measurement. Interestingly, all LPs focused on plane geometry. Only one LP referenced 3D geometry, which was part of the lesson's rationale.

5.2.4 *The development of national lesson plans from the initial tasks to Version 3*

In this section, we return to the national LPs and investigate how they evolved throughout the Intervention, from Version 1 to Version 3 (see Figure 4.1, in Section 4.1.2), and the possible causes of this evolution.

For each LP, the PMTs selected tasks related to R&P. Naturally, these tasks were subject to many modifications due to the assignments before and during Sessions 1 and 2. Some tasks originally selected were later replaced by others; the ones finally used were made more specific and presumably understandable to pupils, and complemented by other tasks to fulfil the intended aims and objectives. They were also embedded into more elaborate implementation instructions. For example, while originally "only" confirming a correct answer in the category "Responding to pupils' reasoning", the Norwegian PMTs suggested "re-represent pupils' thinking and prompting error correction" in response to the Session 2 discussion in which teacher moves were clarified and discussed. Such modifications were seen in all the plans and, in a way, are to be expected considering the deliberate choice of assignments and work within the sessions. The following section will investigate the changes between Versions 1 and 2 and Versions 2 and 3.

5.2.4.1 Modifications between Version 1 and 2

As earlier noted, the Czech PMTs, whose plan was later selected for the Student Meeting, completely changed the content of their lesson after discussing their first ideas in Session 2.[6] Thus, the following text only concerns the four remaining countries.

Through the document analysis, we determined modifications in the plan, which we summarised in tables structured according to the lesson template. To illustrate, we will present an example of the Norwegian plan (Table 5.1). The empty lines show no modification in this part of the template from Version 1 to Version 2. There were some other modifications made that are not apparent in the table, such as a mere rearrangement of the text, a different structuring of the text, and others that did not lead to a substantial change in the LP.

To determine the likely causes of modifications in the plan, we used the analysis of the course of Session 3 to determine what aspects of the plan the PMTs discussed and how the teacher educator led the discussion. If the modification found in Version 2 of the LP could be clearly attributed to what happened in the session, it was written in the second (Peer feedback) or third (Teacher educator's feedback) columns of the table. If not, then it was likely that the modification was a result of the discussion among the authors of the LP before submitting Version 2. Such a modification is presented in the last column of the table.

Our analysis of the summary tables revealed that most modifications were motivated by the discussion among the PMTs in all the plans except for the Turkish one, where the teacher educator's feedback was more prominent.

In particular, the modifications in the Norwegian plan related to the formulation of the learning goal of the lesson, the way the lesson was to be launched, and the homework task, to better connect it with the content of the lesson (Table 5.1).

For the Italian plan, which concerned five lessons, the formulation of tasks was substantially changed to align with the school grades, further improving the lessons' aim. The lessons were specified in timing (the timing of Launch, Main, and Closure activities). Despite the teacher educator's feedback on their importance in lesson planning, anticipated strategies and pupils' misconceptions and difficulties sections remained unelaborated, maybe because of PMTs' low didactic competence. The modifications were probably due to the authors discussing the plan among themselves after the session before submitting Version 2. It is worth noting that the authors of the Italian plan

TABLE 5.1 Modifications in the Norwegian plan from Version 1 to Version 2 and likely causes

Category/ Likely cause	*Peer feedback*	*Teacher educator's feedback*	*Work in the group of authors*
Aims and objectives	Single, clear learning goal, easier for pupils to understand: "Describe figural patterns with your own words"	Learning goal changed, with a closer focus on R&P	Learning goal better aligned with the learning activities
Content			
Learning activities	Modified launch of the lesson, emphasising clarity to make the lesson more accessible and focused for pupils		
Teacher and pupil roles		One of the misconceptions moved among the difficulties	New strategy (creating a table with values) was added to help pupils identify the pattern More detailed guidance provided for the teacher to address misconceptions and anticipated difficulties
Material and resources			
Grouping			Added information on how pupils should work if they have no partner
Location			
Time			
Assessment	Type of question related to self-assessment changed in the exit card		
Homework		Tasks changed so that they align better with the tasks pupils worked on during the lesson	Question added to compare figural patterns in homework to what was done during the lesson
Outline for the next lessons	New activity added (drawing patterns related to a given formula)	Detailed descriptions of tasks for each lesson added Discussion about two patterns added More information ensuring that pupils build on their knowledge and skills systematically	Specific aim for each lesson added

made very few modifications compared to the number of changes made by the authors of LPs in the other countries.

In the Slovak case, the discussion of the LP's authors with their peers uncovered a mismatch between the aims and objectives and the content of the lesson, resulting in substantial modifications in both. Consequently, a new homework task was added, as the previous one was considered trivial. Moreover, more possible errors made by pupils and a description of possible teacher moves were added.

The modifications in the Turkish plan were mostly initiated by the teacher educators' feedback during Session 2 and affected the description of the teacher and pupils' roles most. More information was added on what misconceptions might appear in pupils' reasoning and how the teacher should conduct the lesson (e.g., adding guiding questions).

Overall, we can conclude that the discussion with their peers and the teacher educator in Session 2 influenced the quality of the LP in all the countries, including the Czech one (which led to a complete change of the content).

5.2.4.2 Changes in the plans after the enactment (between Version 2 and Version 3)

There were 36 modifications in total in all the LPs from Version 2 to Version 3, that is, after the enactment of the lesson and its reflection. Half were real changes, while the other half only specified the original plan.

Figure 5.9 shows that the Czech LP received the most modifications, followed by the Italian and Norwegian ones. It is worth noting that only "real changes" were made in the Turkish LP; its authors probably felt that no specifications were needed. Considering that the Italian LP consisted of five lessons and the Slovak plan of two, it is noteworthy that the reflection of the videos did not inspire many changes. Naturally, one explanation would be that the Italian and Slovak LPs were very well elaborated, and everything went according to plan during the enactment.[7] There might be another explanation considering the implementation of the Intervention in Italy (Table 4.2 in Section 4.2). The Italian PMTs did not see the videos before Session 4 and reflected on them in the whole group onsite. Moreover, they had to concentrate on five videos, not one, and some parts of them, which were felt not to deliver much information (such as periods in which pupils were working in GeoGebra), were skipped due to a lack of time. This might have contributed to a rather shallow group reflection, which was not as detailed as for the Czech and Norwegian PMTs, who saw the videos in advance and were asked to conduct individual reflections. In the Czech Session 4, for example, the PMT who taught the lesson started with some observations, and the other PMTs built on it and provided more details from their written reflections. In

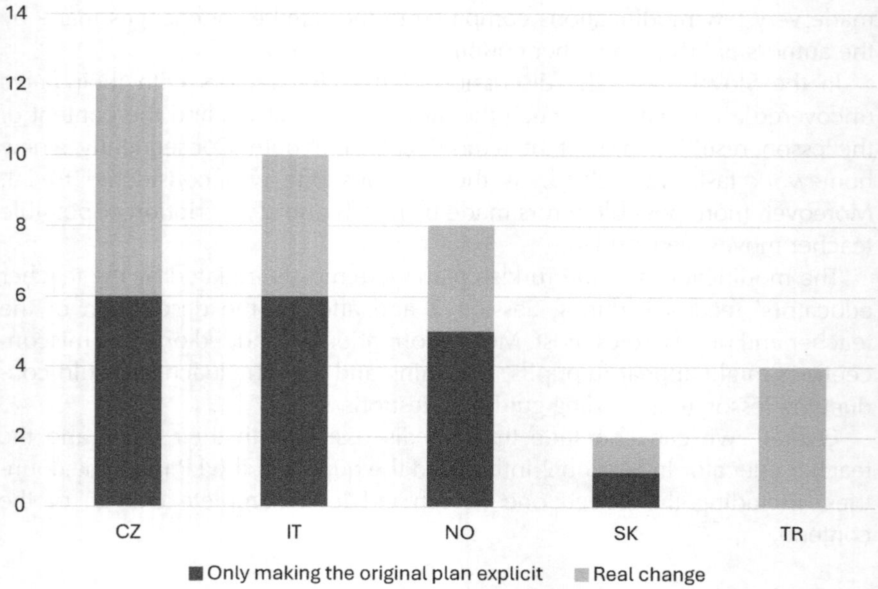

FIGURE 5.9 Number of modifications per country.

the Turkish context, the PMTs also did not see the video in advance, but had more time to reflect on it, as they saw only one lesson.

Let's consider the types of modifications in the LPs according to the lesson planning aspects. We can see that the vast majority were suggested for the learning activities and aims and objectives. The graph in Figure 5.10 shows that PMTs did not change the aims and objectives of their lessons and only made them more specific, but they changed learning activities. Half of the total number of real changes were made in learning activities, thus warranting our attention (four changes in the Czech plan, two in the Italian and Norwegian plans, and one in the Turkish plan).

First, in the Czech LP, the changes in the learning activities were caused by the PMTs not feeling that they sufficiently contributed to the aims and objectives of the lesson. They made a major redesign of Part 3 of their worksheet. In Version 2 of the plan, the pupils were presented with some expressions (e.g., $c^2, 2(e+f)$) and asked to illustrate them geometrically. After the lesson enactment, the PMT who was the teacher and one of the LP's authors realised that Part 3 was not well connected with the rest of the lesson and with R&P. Some other PMTs noted during Session 4 that pupils saw the tasks in the worksheet as independent and that the teacher missed opportunities in the lesson to uncover the connections among them. Not all the PMTs agreed that the original Part 3 was not well connected to the rest of the plan, so the group

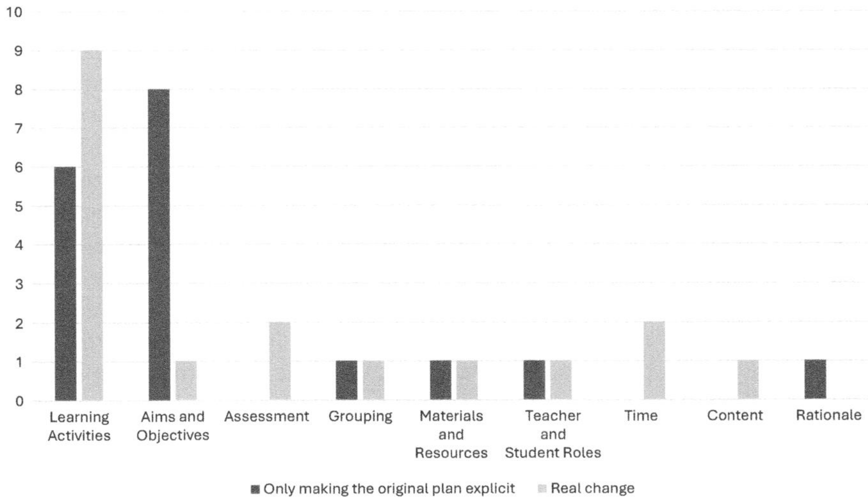

FIGURE 5.10 The number of modifications per lesson planning aspect.

discussed this issue repeatedly during Session 4. Finally, the group agreed on the change. In the original Part 3 of the worksheets, the tasks were not connected to Part 2, in which algebraic expressions were modified, and pupils should check the correctness of such modifications. The pupils fulfilled the tasks by referring to algebraic formulas without being encouraged to provide other arguments, such as geometric ones. As before, the new version, Part 3, asks to provide a geometric picture for an algebraic expression (e.g., $(a-b)^2$) but now, vice versa, also requires pupils to write an algebraic expression for given pictures (e.g., for a rectangle with sides of a and 3, or a right-angled triangle with two sides of length b). Compared to the previous version, what is also new is that the algebraic expressions in Part 3, with their geometric representation, were selected so that they can be used to show the incorrectness of the modifications presented in Part 2.

Second, the Italian PMTs added two more tasks to their plan – one was actually used by the teacher during the lesson, and the PMTs appreciated it because of its clarity and the possibility of integrating aspects of content not previously considered. The second elaborated on one of the tasks already in the plan. The PMTs paid more attention to it in terms of the language used.

Third, the Norwegian PMTs changed the way the main tasks were presented in the worksheet. They suggested more targeted questions and separated them graphically. Some tasks were formulated in slightly different ways, which they felt could influence the quality of classroom discussion in terms of R&P. They also specified how the tasks should be introduced to the pupils, emphasising justifications.

Finally, the Turkish PMTs added another objective (focusing on the identity $(a-b)^2$), an extra task, and expected correct and incorrect responses for this task, as there was time left during the enactment of the lesson, but otherwise left the content unchanged. They also revised the short-term assessment to align better with the lesson content (showing the identities using modelling).

To mention some other noteworthy modifications, in Session 4, the Slovak PMTs said they did not like how the cooperating teacher enacted the lesson. They felt that she was too directive and did not provide enough space for the pupils, and they realised that more information about the required socio-mathematical norms was needed in the LP. In the original version, the section on norms only included tasks. The added paragraph includes norms related to the teacher and pupil roles. The two changes in the assessment depicted in Figure 5.9 consist of Italian PMTs suggesting slightly different questions used for assessment, with more open questions and more understandable wording, and Turkish PMTs using a new, more explicit task. Next, the Norwegian PMTs realised that pupils spent a lot of time squaring two-digit numbers, which was not the focus of the lesson, and decided to allow the use of a calculator to provide pupils with more space for reasoning and justification. Finally, the Czech PMTs struggled to formulate the aims in relation to the age group of pupils. For example, the original aim was to "define what a proof is", even though the PMTs wanted the pupils to understand that justification in mathematics is important and that not all justifications (such as empirical ones) are proofs. The inappropriate formulations were changed after discussing the lesson video in Session 4, in which they could observe what the pupils were actually learning.

5.3 PMTs' experience of the Intervention

Section 4.1 provided the rationale for the design features built into the Intervention. After exploring the PMTs' beliefs concerning R&P in Section 5.1, we now focus on their perception of the Intervention set-up. The analysis of the data presented in Table 4.4 in Section 4.3 yielded results presented separately for each design feature in Sections 5.3.1 to 5.3.3. The results of the analysis of how the PMTs participating in the International Student Meeting perceived the set-up of this meeting and its affordances are presented in Section 5.3.4.

5.3.1 PMTs' reflections on text resources

In their Final Assignment, many PMTs from each country commented on the reading part of the Intervention. Let us reiterate that the five texts assigned to the PMTs during the Intervention were modes of reasoning, teacher moves, spider web, and socio-mathematical texts (two texts). They are presented in Section 4.1.1

in the order of "appearance" in the Intervention. In this section, we present them in the order of "popularity" among the PMTs identified in the data.

5.3.1.1 Spider web

Akker's (2013) "spider web" was generally seen positively as a tool helping PMTs in lesson planning and ensuring that all essential elements were considered in the plan:

> When making teaching materials, Akker's spider web is very useful for me as it forces me to think about the tasks, their order and forms of teaching from various aspects which directly influence the quality of acquired knowledge.
>
> *(SK)*

Even Norwegian PMTs, who were skilled in lesson planning, since before the Intervention they conducted 95 days of practice placements, and Turkish PMTs, who started lesson planning early in their studies, found the spider web beneficial, as it required them to think about details, reflect on their choices, and plan ahead:

> I found it very interesting to use the Spider Web model. Previously, I have used the didactic relational model and the classic what/how/why table in planning, and over time, one might become less aware of the components that are actually included in these, as it becomes somewhat automatic. Using a new model made me step back and reflect more on my own choices, and I felt that I developed my own competence in lesson planning as a result of this.
>
> *(NO)*

> Thanks to this [template], we planned the next lesson, at least we made a draft of it. You know, I think it was also very advantageous in terms of the annual plan. You know, planning a few lessons ahead, I mean, more, how should I say, moving forward with a plan, you know, it was good for me in that respect.
>
> *(TR)*

Interestingly, some Norwegian PMTs reported difficulty following the spider web terms in English in the exit cards.

5.3.1.2 Teacher moves

Ellis et al.'s (2019) text on teacher moves also met a widely positive response ("For me, what we studied about teachers' moves in the classroom and task roles, in promoting and assessing R&P competencies, was illuminating. I have never had the same input!", IT). The PMTs were often surprised by the perceived power of teacher moves to influence the depth of pupils' understanding of mathematics:

> [The topic of teacher moves] seems to me extremely useful because I think that by very small changes, for example, using a few different words, a teacher can make a big difference in what pupils get from the interaction with the teacher.
>
> *(SK)*

> [...] perhaps I had previously, consciously or unconsciously, thought that asking 'why do you think that' or 'can you explain what you were thinking' were the main ways a teacher could promote pupil reasoning. The introduction of this framework, and the work in lectures and assignments, has given me a deeper understanding not only of what a teacher can do to promote pupil reasoning but also of the significance of these actions.
>
> *(NO)*

> Well, you know, these teacher moves, especially, impressed me a lot. I talk about it all the time, because before it felt like: "If you provide a discussion environment with the child, it is enough" [...] But you know, in those teacher moves, low-level discussions can be provided, high-level discussions can be provided. I became aware of this.
>
> *(TR)*

One PMT appreciated the role of teacher moves in designing teaching specifically aimed at R&P:

> This article [on teacher moves] also opened up the most for me about how reasoning and proving should be done in teaching. The examples in it were clear and illustrative. I reflected on this article when planning our lessons and tried to use it to create questions that supported pupils' reasoning. [...] I have always tried to encourage pupils to explain their opinions, usually by asking questions: Why? However, the article offered many other options to target when we want the pupil to justify their answer, or perhaps even to prove a point.
>
> *(CZ)*

Notably, some PMTs understood teacher moves only later when reflecting on the enactment of their LP. Some PMTs would welcome more time devoted to teacher moves, as they felt they could not understand them well enough:

> The different categories by Ellis et al. have made me aware of how much a teacher can do, while also showing me that I need to delve even deeper into them before I can say that I have sufficient knowledge of the framework.
>
> *(NO)*

Unlike the Final Assignment, exit cards captured PMTs' views immediately after the session and highlighted some problems with understanding the assigned texts. Even though generally very well accepted, the distinction between teacher moves was difficult for some PMTs, as "some of the teacher moves are very similar to each other" (TR).

5.3.1.3 Modes of reasoning

Another frequently mentioned problem in all the countries was distinguishing the modes of reasoning introduced by Silverman and Even's (2015) text. Without discussing them in the sessions, the PMTs would not understand them; even after that, some were unsure about them.

> After each group presented their task, the discussion about the reasoning mode of the task was the best part of the lesson for me because I had the opportunity to think about points that I had never thought of during these discussions. In addition, the points that I did not understand or confused became a little clearer thanks to these discussions.
>
> *(TR)*

> I liked the examples of reasoning. Because even in my homework, I didn't decide truly what modes of reasoning are used in my task. This part really helped.
>
> *(TR)*

In the final interviews, some Czech PMTs considered it important to know about different forms of reasoning; however, they did not see the distinction as important, as the categories overlap with each other. Some PMTs highlighted that they needed time to make sense of modes of reasoning.

> When we started with reasoning and Silverman and Even, I noticed that I was unable to distinguish between the forms of reasoning the pupils were using. This was probably the part of the MaTek project that I found most difficult. However, by the end of the project, I am now much better at distinguishing between the different forms of reasoning the pupils use.
>
> *(NO)*

Some PMTs thought that they understood the modes of reasoning well, but were reluctant about how they should teach them.

> I think I can now, in many cases, fairly easily categorise and identify the different ways pupils reason. What I would like to work more on is how I can help pupils develop this even further. For example, how can I help pupils move from empirical reasoning to deductive reasoning? ... I would like to work more on how I can do this, not just in the steps I take during the lesson, but also in how I structure the sessions and what kind of tasks I present to the pupils.
>
> *(NO)*

5.3.1.4 Socio-mathematical norms

The text on socio-mathematical norms was rarely mentioned in the data. Some Czech and Italian PMTs found the part on socio-mathematical norms difficult to read in English. One text is part of an encyclopaedia, a condensed professional text with many references and few examples. It seems to be too difficult for PMTs.

On the other hand, some exit cards show that PMTs appreciated the discussion about socio-mathematical norms in the session.

> I learned that norms actually allow pupils to express themselves, the teacher needs to make-an-effort to create the norms, and then the norms can turn into a routine.
>
> *(TR)*

> I really liked the sentence that mathematics determines what is true. That in proving something, it's not about who expresses themselves better or who has a stronger opinion. Mathematics is quite unique in that it is universal. It's really just about whether something is true or not. And I found that to be a nice reflection on the nature of proof.
>
> *(CZ)*

Thus, it might be more appropriate to intuitively introduce the concept of norms using examples and discussing it onsite before assigning the text to read.

5.3.2 PMTs' reflections on task design and lesson planning

PMTs often commented on task design itself, without referring to lesson planning. They mostly expressed surprise that "there is so much behind writing a task" (IT) or "I therefore liked experiencing how much small changes to tasks can actually impact pupils' learning" (NO), and that one task could be modified in many ways, including making the task a rich one.

> It was equally useful for me when we redesigned tasks, when we changed a low-cognitive demand task into a higher-cognitive demand task or when we reformulated them as good questions. It was equally useful that we did not only talk about these things but we were to realise/create them in practice.
>
> *(SK)*

After experiencing lesson planning, some mentioned that being an R&P task is not an innate quality: "Many tasks could be redesigned to be a reasoning and proving task." (CZ)

Not surprisingly, many PMTs felt in their exit cards that redesigning and formulating tasks was difficult for them and that they needed more experience to feel at ease. Some were confident in their final appraisal:

> I feel it has become easier to critically assess tasks given in textbooks. I believe that after these sessions, I will be better at evaluating the tasks I plan to give to the pupils, to see if they actually make sense and if they promote the focus I want to emphasise in the lesson.
>
> *(NO)*

Lesson planning was met with a very positive response from all the countries. PMTs expressed in different words that the "spider web" helped them see the surprising depth of thought and detail they must consider to plan a lesson.

> From identifying the learning objective to choosing tasks, activities, etc., every aspect of the lesson requires careful thought and planning. I have learnt a lot about this that I wouldn't have expected before the intervention. Knowledge of mathematics is a necessity for teaching, but it is not enough.
>
> *(IT)*

In their exit cards, similarly to the task design described earlier, many PMTs expressed their worries about their planning capacity (mostly referring to the fact that it was a fairly new experience for them) in general and planning for R&P in particular. In the case of Norwegian PMTs, who were used to planning lessons, some appreciated the spider web template as documented in Section 5.3.1. In contrast, others expressed difficulty in using the spider web template because "it is an unfamiliar tool" or "it is larger and more elaborate than I'm used to".

The lesson planning experience provided more meaning to the concepts previously introduced by reading, as illustrated by the quotes from the final interview. Such PMTs appreciated the coherence of work in the Intervention.

> At first, we didn't really know what we were going to do, because we had just started reading the articles. But as we read the articles, we said, "How can we use teacher moves?" or "How can we improve those reasonings?" In that way, since we did it in a certain period of time, I think we improved on it every week, every week. And we developed our lesson plan accordingly.
>
> *(TR)*

Working in groups was fairly common for PMTs, except for Italian PMTs: "Working in a group was a new experience for me. I had hardly ever worked like this with my colleagues in the classroom during my studies." (IT) PMTs in all the countries, both in the exit cards and in the Final Assignment, appreciated collaborative lesson planning in groups where there were different ideas about the lesson and in groups where they fully agreed.

> I found the lesson planning in groups appropriate. In the group, we met with like-minded colleagues and had a common idea of how we wanted to design the lesson.
>
> *(CZ)*

> Working in groups suited me especially because we shared different opinions and ideas with each other. We could consult them among ourselves and then try to come up with a version that we all found ideal. I found it interesting to discover that although we were preparing the same teaching unit, each of us had a slightly different idea of how to teach it, and the teaching would end up looking slightly different from each of us.
>
> *(CZ)*

The group work helped those who were not sure about their capacity to plan:

> It is also beneficial that when you present your work, you are not alone in it. If you need to defend some aspects of your work, it helps that other members of the group add arguments and you explain it together what your intentions were.
>
> *(SK)*

> The fact that we had a groupmate who is currently teaching helped us a lot in terms of how pupils might make mistakes and what difficulties they might experience while explaining this topic.
>
> *(TR)*

The PMTs mainly appreciated the discussion when preparing the plan and could feel that such discussions helped them learn ("The discussion [...] is always a source of inspiration for me", SK).

There were a few negative comments about group work. One aspect concerned time – for some PMTs, it was too time-consuming, and it was difficult to find a common time for work. Second, the disparity of opinions on the content of the lesson and how it should be delivered was seen as a hindrance, as "it was not easy to reconcile and agree on common results". (SK) Another PMT was concerned that it was difficult to share the work in such a way that "each made approximately the same part". (SK) The same PMT complained that they often had differing views, leading to misunderstanding criticism. Another PMT clarified that the cause of such problems was their differing views of teaching mathematics. However, no such comment appeared in the Norwegian and Turkish data, as group work is customary in the PMTs' lessons.

5.3.3 PMTs' reflections on lesson plans and on the enactment of the lesson/s

In the Final Assignments, most PMTs spoke about how they prepared the lesson, not how they reflected on the plan of the other groups or the enacted lesson. It seems that the concept of reflection did not stand out to them as a specific feature of the Intervention but was seen as something usual. Indeed, the PMTs in all the involved countries are often asked to consider and comment on the work of their peers. The reflection on the enacted lesson was a one-time thing in the Intervention, which may be why they did not single it out for mention.

There were a few exceptions. For example, the following quote shows that this PMT appreciated the reflection as a way to judge the decisions made during the planning stage:

> After observing the lesson while it was happening and after watching the video recordings, I am very glad that we came to this conclusion [changing the individual work to work in pairs]. My point is that it has been very useful and instructive to be able to reflect on the choices made in planning the lesson and to receive feedback on what has been done. This forces you to think about the decisions that may have initially been a bit random or made with a different 'lens' than the one you become aware of over time. *(NO)*

Similarly, two PMTs, one Norwegian and one Turkish, who were the authors of the enacted lesson, appreciated being in the class and later also discussing the video in the session. This apparently helped them to see the power of teacher moves and well-thought-out tasks, respectively:

> I experienced the [potential of teacher moves] after observing [teacher name]'s lesson – both when we attended and observed the lesson, and when we participated in discussing the video in class. I found it very useful to see how the strategies [teacher name] used during the lesson contributed to more reasoning than I believe the pupils would have achieved on their own.
>
> *(NO)*

> I didn't know that pupils were so open to thinking. Some pupils can really, uh, when we give them a good thinking time and when we give them the necessary materials, they can really do proof. They can defend their thoughts. They can explain where they came from and how they came to their conclusions. I was surprised by this, to be honest.
>
> *(TR)*

What appeared more often was the concept of feedback from others on their own work. For example, Turkish and Norwegian PMTs explicitly mentioned the help of the feedback received in improving/revising the LPs from their peers and instructors during the sessions.

> While preparing our own lesson plan and modifying it according to the feedback we received, as well as listening to the plans of our friends in other groups, I had an idea about how reasoning and proof could be used in the math lessons we prepared.
>
> *(TR)*

> The presentation and feedback on the lesson plan were interesting because, as a group, we felt that we had a specific view on our task and what it was supposed to promote during the lesson, and we felt that this was aligned. However, in the feedback from the other groups, we discovered that they had a different perspective on the tasks that we had not considered, which meant the task might not necessarily promote reasoning in the way we thought.
>
> *(NO)*

In the exit cards, which responded to the immediate work in the sessions, some PMTs mentioned that they liked being given feedback on their LP, including the negative one.

> It was important for me and my colleagues to receive criticism and/or praise when presenting our plan.
>
> *(IT)*

> Which part of the session did you like best today? I liked receiving criticism on your own submission, and it's fine to get suggestions for improvement.
>
> *(NO)*

> Which part of the session did you like best today? 'Roast' of our plan (feedback).
>
> *(NO)*

Some appreciated the feedback from the teacher educators on their plans: "The constant feedback from our instructors and the interaction between everyone in the classroom broadened my perspective on the lesson." (TR) Some Czech PMTs would appreciate more concrete feedback on their LP. The teacher educator provided oral feedback during the sessions as part of the discussion with the PMTs; the PMTs might have expected written feedback (as is customary for some assignments in the mathematics education course of the Czech PMTs).

The Norwegian PMTs explicitly valued the reflection on the enactment of the lesson they designed as a fitting closure of the Intervention, as it embedded all the features of the Intervention:

> I kind of feel like the video we watched was, in a way, a sort of, shall we call it a summary, of everything we've gone through. Because we got to see both the pupils' reasoning and proof, various moves the teacher used, and the design of a lesson plan. So, it was like, what should we call it, a nice overview, or you got to see everything we had worked on throughout the semester.
>
> *(NO)*

> It was a really good summarising task or activity, watching the [enacted] lesson and then discussing with the group afterwards. And I felt that it was very relevant to the profession, in the sense that we could sit and discuss things we don't really know much about yet.
>
> *(NO)*

Even though the PMTs sparsely commented on the aspects of the Intervention related to reflecting on the LP and its enactment, we can indirectly infer how these aspects influenced their lesson planning. In Section 5.3.3, we pointed out how discussing the LP with their peers and with the teacher educator in Session 2, and after its enactment in Session 4, influenced the authors' subsequent elaborations of the plans.

5.3.4 PMTs' reflection on the International Student Meeting

A majority (21/30) of participating PMTs stated that they most appreciated having worked in mixed/international groups during the meeting in Prague. The most commonly mentioned affordance of this international collaboration was the exposure to diverse perspectives and teaching approaches. Participants frequently highlighted how working with individuals from different nationalities and educational systems broadened their understanding, allowed them to compare and discuss varying methods, and provided insights into how lessons can be planned and executed differently. For example, an Italian PMT reflected: "[…] it was very interesting to try to create a lesson from scratch using alternative methods to ours", and another Italian PMT emphasised the different ways reasoning is supported in various educational systems:

> What I consider important is the support that in other countries is given to students in the moment of reasoning, such as trying to make them think for themselves and, for example, arriving at the formula in different but effective ways.
>
> *(IT)*

While mainly Italian PMTs claimed to benefit from gaining new perspectives on teaching R&P, some PMTs highlighted the role of R&P and its importance in mathematics itself. A Slovak PMT explained: "It was not about learning something new, but rather about confirming to myself that reasoning is very important in mathematics." Similarly, a Czech PMT stated: "Even before the meeting, we perceived the importance of reasoning and proving in mathematics lessons. I would say that the meeting made me even more convinced of their importance." Alongside this reinforced importance, some PMTs also expressed a stronger belief regarding the place of R&P in different topics of school mathematics.

> I was once again convinced that every topic offers space for reasoning and proof (even in more challenging topics that we don't address in elementary school – even those can be incorporated into a class curriculum in a simpler form).
>
> *(SK)*

A Czech PMT expressed a similar belief by considering the role of teacher moves in supporting reasoning: "[I learned] that it is possible to implement various teacher moves into many different topics."

The theme of how PMTs conceptualise R&P emerged in their responses. Several Norwegian PMTs reflected on the diversity of perspectives on R&P across nationalities, with one stating: "I was not prepared for the fact that we had such different views on what reasoning actually entails in the different nationalities." Another noted: "[Because of realising] that there are different views on what reasoning is, [...] I gained a lot of confidence in my own competence in the area of reasoning." For some PMTs, the meeting afforded a shift from a procedural to a conceptual understanding of R&P. A Slovak PMT articulated this change: "I learned that if we want to support students' reasoning, we should focus not so much on "How" questions but on "Why" questions. Or not stop at the "How" question, but expand it with a "Why" question. Similarly, the meeting led some Italian PMTs to rethink the role of R&P in mathematics education, shifting their focus towards reasoning as a process that fosters deeper understanding.

> The meeting in Prague made me think that in order to teach a lesson I have to put myself in the shoes of the pupils and be able to answer their possible "whys". It is important to help pupils understand the proof, rather than presenting a proof without reflection.
>
> *(IT)*

Another Italian PMT highlighted how collaborating with peers during the meeting deepened their awareness of the role of questioning in guiding pupils towards proof:

> [...] while we were collaborating, this was precisely the recurring question, the same question a typical math student asks a teacher: "Why should I do this? What is it useful for?" This made me even more aware of how important it is to be able to explain the "why", to encourage reflection, reasoning, and to guide students toward proof.
>
> *(IT)*

Through their participation in the meeting, these PMTs developed a greater appreciation for reasoning as an active, student-centred process, recognising the teacher's role as a facilitator who helps pupils understand and construct proofs themselves, rather than passively observing the teacher's demonstration. The role of the teacher was a recurring theme in the reflections of PMTs from different countries, particularly in the context of lesson planning. For example, a Czech PMT acknowledged the importance of anticipating student difficulties and guiding their discovery process, stating: "In my lessons, I need to improve thinking about the right questions that I will ask the students. I have to have a plan if the students get stuck and don't discover what I want them to." Similarly, a Turkish PMT highlighted the significance of incorporating purposeful questioning into lesson design:

> I learned how to write a lesson plan more practically and quickly. The lesson plans we wrote were longer and more difficult to review. Besides, I learned how to add why and how questions while preparing a task.
>
> *(TR)*

The meeting also afforded a stronger focus on pupil reasoning within LPs.

> The most important thing I learned during the meeting was probably how easy it can be to "forget" pupil reasoning in lesson plans. I notice that during the meeting, I am more and more convinced of this, and it is something I can take with me further in my own practice going forward.
>
> *(NO)*

This awareness aligns with the perspective of another Czech PMT, who recognised the importance of assessing pupils' understanding throughout the lesson, saying: "I learned that I should use exit cards in my lessons and focus on the aims of the lesson. I should check what the pupils learned during the lesson."

Together, these reflections underscore the evolving role of the teacher, not only as a planner of structured lessons but as a facilitator of pupils' reasoning, through the ability to ask the right questions, anticipate difficulties, or continuously assess pupils' understanding.

While the international collaboration was widely appreciated, some PMTs also identified challenges. The most frequently mentioned one was the language barrier. Additionally, some PMTs highlighted differences in the conceptualisation of R&P and approaches to its teaching. For example, a Norwegian PMT observed: "They wanted an approach where the teacher gives instructions, while I wanted a more active student group." Curriculum differences were another challenge, as noted by a Czech PMT: "From the beginning of the discussion about what is taught where and in which year, we could not immediately agree." A Turkish PMT observed variations in the use of digital technology: "Other countries do not use digital tools, it has surprised me because we always apply technology to our lesson plan."

Despite the diversity and challenges in terms of "the different views on the issue" (as noted by a Czech PMT), the Student Meeting and its activities (mainly the collaborative design of LPs in mixed groups but also presentations of national LPs, received feedback, and information about the school system in each country) were seen as enriching and a significant factor in PMTs' learning experience.

Notes

1 Referring to ways of showing pupils why multiplication works and how we can show pi to the pupils experimentally.
2 In the Turkish group, this conception of R&P was also present after the Intervention as some quotes from the post-interviews showed.
3 Not necessarily called that, the word culture was used more often even though the PMTs met the concept of socio-mathematical norms during the Intervention.
4 Hejny method is a constructivist, scheme-oriented way of teaching mathematics used in about 20% of Czech primary and lower-secondary schools. https://www.h-mat.cz/en/hejny-method
5 https://mathigon.org
6 Their first idea was dismissed when discussed during the session as the PMTs (and the teacher educator, too) felt it was not an R&P task at all.
7 In this context, it is noteworthy that after the enacted lesson, the Italian PMTs realised that five lessons are not enough and suggested adding two more lessons. Thus, the analysis of the Italian LP in Section 5.2.1 concerns seven lessons.

6

DISCUSSION OF RESULTS

This chapter presents a critical analysis and interpretation of the findings outlined in the previous chapter, focusing on their implications within the context of mathematics education. Drawing on the study's research questions and theoretical frames, the discussion seeks to connect the empirical results to the existing literature, highlight patterns, and explore possible explanations for the observed outcomes. Particular attention is given to how the results contribute to a deeper understanding of pre-service mathematics teachers' (PMTs') development of lesson design capacity, their understandings of reasoning and proving (R&P) in school mathematics classes, and perceptions of R&P in the curriculum. The chapter also considers the broader educational significance of the findings and identifies areas for future research and practical application.

6.1 PMTs' beliefs of reasoning and proving in teaching before and after the Intervention (RQ1)

Our targeted intervention has demonstrated a significant **transformative impact** on PMTs' beliefs about R&P. Participants, especially those from Slovakia (SK) and Italy (IT), experienced a significant shift in their understanding of R&P. Initially seen as a distinct and isolated topic, R&P came to be viewed as a fundamental element that cuts across all areas of mathematics. In both SK and IT, this shift went further, as PMTs began to appreciate the presence of R&P throughout the curriculum and its informal and exploratory dimensions, moving beyond an exclusive focus on formal proof. These changes reflect findings in the literature that stress the importance of expanding PMTs' views of R&P beyond a purely formalist approach (Buchbinder &

DOI: 10.4324/9781003533870-6

McCrone, 2020; Stylianides & Stylianides, 2013). Moreover, PMTs acknowledged that R&P can be meaningfully integrated into mathematics instruction at all levels of schooling, as long as it is tailored to students' developmental stages and supported by diverse representations, not limited to abstract symbolism. This perspective was particularly noticeable in SK and the Czech Republic (CZ), possibly influenced by their shared educational and historical backgrounds.

The Italian PMTs in our study highlighted geometry as the most suitable topic for R&P, a perspective rooted in their prior experiences as university students of mathematics. This finding aligns with Dickerson and Doerr's (2014) observations that teachers' initial beliefs about proof are often shaped by their own educational experiences, which may lead to a narrower understanding of its scope. This shift underscores the transformative potential of exposing PMTs to diverse contexts and innovative practices, as Cai (2025) and Boesen et al. (2010) emphasised. Such experiences challenge traditional conceptions of R&P as confined to geometry, highlighting its broader applicability across the mathematics curriculum. This evolution in perspective also reflects Michal's (2023) findings for Czech PMTs on the importance of presenting pre-service teachers with varied representations to foster a more integrated understanding of R&P.

In the pre-interviews, PMTs primarily described R&P as the justification of solutions, including commenting on steps, procedures, or theorems. However, justifying the correctness of individual steps was notably absent in the post-interviews. Some PMTs in the pre-interviews also interpreted R&P as making connections within mathematics, aligning with the understanding that mathematics is the "glue" that holds mathematical knowledge together, a framework proposed by Kilpatrick et al. (2001).

A clear distinction between reasoning and proving emerged in the TR and IT groups. We contend that this was likely to be influenced by their educational backgrounds: Italian PMTs had a strong mathematics foundation in their courses, which emphasised proof as the backbone of mathematics. At the same time, they had no possibilities of designing lessons and even less of teaching them. Turkish PMTs had multiple opportunities to prepare lessons in their courses. PMTs' emphasis on "explaining it to someone" and using "their own words" suggests that they viewed reasoning as the process of sharing and communicating their thought processes. This conceptualisation resonates with Kilpatrick et al.'s (2001) assertion that explaining or justifying solutions to others is critical for developing reasoning skills. Notably, in the NO groups, R&P was often framed as the process of communicating mathematical thinking, which aligns with Herbert et al.'s (2015) research. Across groups, during the Intervention (between pre- and post-interviews), many PMTs moved away from viewing R&P solely as the construction of rigorous proofs. This is in line with the findings of Dickerson and Doerr (2014) that less experienced teachers often held the view that proofs should conform to rigorous standards of language precision and logical reasoning.

Post-interviews revealed several roles emphasised by PMTs, including justifying the existence and validity of mathematical knowledge, fostering pupil engagement, facilitating explanations, promoting deep understanding over rote memorisation, and emphasising convincing and effective communication. Compared to the pre-interviews, PMTs expressed these roles with greater conviction, describing the Intervention as transformative and recognising R&P as critical components of mathematics education.

These findings align with recent research on pre-service mathematics teachers' perceptions and the impact of interventions such as ours. For example, Boesen et al. (2010) emphasise the significance of tasks that promote reasoning and deep understanding, while research by Stylianides (2009) highlights the transformative potential of R&P interventions in reshaping teachers' understanding of effective mathematics teaching. Moreover, studies such as Cihan and Akkoç (2019) report that interventions targeting PMTs' R&P skills not only enhance their ability to engage learners but also help them perceive R&P as essential in developing pupils' mathematical thinking and justification skills. These findings further emphasise the importance of interventions that encourage effective communication, deeper conceptual engagement, and a shift from rote memorisation.

The **concerns** raised by PMTs in the pre-interviews, such as challenges related to curriculum demands, time constraints, the perceived difficulty of R&P for pupils, the absence of socio-mathematical norms in classrooms, and a lack of prior experience with R&P, remained present but were expressed differently in the post-interviews. Time constraints were a persistent concern. Initially, PMTs feared R&P would slow lesson pace, especially in mixed-ability classes. Later, concerns deepened: planning R&P, ensuring broad student engagement, and building supportive norms all required time. These findings reflect the tension between efficient instruction and reasoning-rich teaching (Inglis & Alcock, 2012; Jones, 2010). Some Czech and Slovak PMTs doubted whether this time investment would lead to visible gains, echoing concerns under curriculum pressure (Cirillo, 2018). Interestingly, some Turkish PMTs no longer perceived time as a limitation after implementing the R&P lesson, suggesting that practical engagement can reshape perceptions of feasibility. This finding aligns with Boaler's (2022) emphasis on how exposure to innovative teaching approaches can transform teachers' beliefs about the accessibility of complex mathematical practices.

Post-interviews revealed a stronger focus on **the teacher's role** in shaping classroom norms that support R&P. PMTs recognised that their interactions with students are central to establishing socio-mathematical norms – shared expectations that make reasoning, justification, and proof integral to discourse (Yackel & Cobb, 1996; Kilpatrick et al., 2001). Although R&P was still considered challenging, the reasoning shifted: it was no longer considered suitable only for talented students. This reflects a broader understanding of its accessibility, aligning with Stylianides and Stylianides (2017), who argue that R&P can be scaffolded for diverse learners, and Boaler (2022), who promotes equitable access to challenging mathematical tasks.

Before the Intervention, PMTs often attributed students' struggles not to cognitive limits but to prior experiences of mathematics as rule-based rather than exploratory. This supports findings that socio-mathematical norms are key to fostering argumentation- and proof-related discourse (Stephan & Akyuz, 2012). Yet shifting these norms proved difficult. After the Intervention, many PMTs noted that students were unfamiliar with reasoning in mathematics lessons and needed explicit, structured support. They increasingly acknowledged the teacher's responsibility in cultivating a reasoning-oriented classroom through questioning and interaction, rather than expecting such norms to develop independently (Stephan, 2020).

Our findings align with those of Cai (2025), who demonstrated that reflective practices during professional development help teachers appreciate the interconnected nature of R&P across mathematical topics. Similarly, in our study, PMTs benefited from structured opportunities to reflect on their teaching, which may account for some regional differences observed. Dickerson and Doerr (2014) argue for moving beyond rigid, formal conceptions of proof towards more flexible understandings that include diverse representations – a key realisation among our participants. While some initially saw R&P as a skill suited only for advanced learners, many came to view it as a fundamental component of mathematical thinking that should permeate everyday instruction. Several PMTs even described the Intervention as an "eye-opener" in this regard. This shift mirrors recent curricular reforms, such as Norway's LK20 (Utdanningsdirektoratet, 2020) and Slovakia's forthcoming 2026 reform (MinEdu, 2023), both of which place reasoning and argumentation at the core of mathematics education. These reforms reflect a broader international trend emphasising R&P as essential for fostering deep understanding, critical thinking, and problem-solving (Niss & Højgaard, 2019; OECD, 2018). The growing curricular emphasis underscores the relevance of our findings, particularly the challenges PMTs foresee in translating these expectations into practice. Teachers' ability to integrate R&P into their instructional design is pivotal, as highlighted in studies such as Stylianides and Stylianides (2013), which explore how reasoning and proof foster deeper mathematical understanding. Similarly, the work of Reid and Knipping (2010) demonstrates how opportunities to learn reasoning and proof shape pre-service teachers' capacity to design tasks that support student exploration and argumentation. Crucially, perceptions of R&P often influence how it is implemented in the classroom. Durand-Guerrier and Tanguay (2018) point out how teachers conceptualise proof, ranging from formal logic to exploratory processes, underscoring the importance of shaping pre-service teachers' understanding early in their training. These perspectives align with the Intervention's emphasis on developing task and lesson design skills that prioritise reasoning and evidence-based argumentation.

These studies highlight the transformative potential of targeted interventions, reflective practice, and tailored support in reshaping PMTs' beliefs about R&P and its role in mathematics education.

6.2 Lesson plans and their developments (RQ2)

Our findings support research suggesting that different national traditions shape how reasoning and proof are integrated into school mathematics (Stylianides & Stylianides, 2009). In some systems, proof is treated as an advanced, formal process, while in others, it is introduced gradually through conjecturing, exploration, and argumentation (Hanna & de Villiers, 2012). The SK and TR cases illustrate how an emphasis on student exploration can sometimes lead to reasoning being seen primarily as empirical verification rather than formal deduction (Knuth, 2002).

Across all groups, a pupil-centred approach was prioritised, reflecting a shared belief in fostering active student engagement (Schoenfeld, 2010; Strachota et al., 2018). This approach aligns with constructivist theories (e.g., Piaget, Vygotsky) and research emphasising active learning as a pathway to deep understanding (e.g., Boaler, 2022). Teachers were positioned as facilitators, guiding pupils through questioning and scaffolding, a role widely supported in mathematics education literature (Anthony & Walshaw, 2009; Stein et al., 2008). However, the depth of reasoning expected from pupils and the explicit articulation of reasoning paths varied significantly across groups, even though PMTs demonstrated a strong commitment to promoting student participation and valuing multiple solution strategies. This suggests that PMTs may need additional support in embedding reasoning deliberately and confidently into their lesson planning.

Looking at the evolution of the national LPs from their initial versions to the final adaptations (revised after enactment), it becomes evident that while many aspects of the plans underwent several changes, R&P remained largely "untouched" across all cases. Despite the explicit assignment of designing and enacting a lesson that fosters R&P, its presence was often implicit rather than explicitly structured within the plans. While PMTs made various modifications to instructional strategies, assessment methods, and classroom activities based on their enactment experiences, the way reasoning was incorporated remained largely unchanged. Based on the interviews with PMTs, they recognised the importance of R&P but struggled to operationalise it within their LPs, reinforcing the well-documented challenge of explicitly integrating reasoning into teaching practices (Stylianides & Stylianides, 2009; Lithner, 2008).

Research has shown that teachers often revert to procedural approaches when faced with classroom constraints (Inglis & Alcock, 2012; Hanna & de Villiers, 2012) and that R&P is frequently perceived as an isolated component rather than an integral aspect of mathematical activity (Pedemonte, 2007; Durand-Guerrier et al., 2012). The persistent difficulty in embedding reasoning within lesson structures suggests a need for more explicit scaffolding and professional development to help PMTs transition from theoretical appreciation to practical implementation.

Several key patterns and contrasts emerge when comparing the LPs from international groups with those developed in national settings. Both national and international groups emphasised pupil-centred instruction, which is consistent with constructivist perspectives on mathematics teaching (Boero, 2007; Lester & Cai, 2016). Experimental demonstration was a recurring reasoning mode, serving as a bridge towards formal reasoning, though its role was sometimes left implicit. The teacher was envisioned as a facilitator rather than a direct instructor, aligning with research on effective R&P instruction (Evans et al., 2022). Most groups aimed to help pupils transition from empirical reasoning to deductive reasoning, yet explicit planning for such transitions was often missing (cf. Pedemonte, 2007).

The inclusion of real-world contexts in some LPs reflects efforts to make mathematics relevant and engaging. This practice resonates with research emphasising the motivational and contextual value of real-world problems (English & Gainsburg, 2016; Gravemeijer & Doorman, 1999). However, PMTs primarily used these contexts for motivational purposes rather than as a basis for fostering reasoning. Future training could focus on leveraging real-world problems to drive reasoning and proof more effectively. An innovative approach was seen in Group 1's lesson, which planned outdoor activities as the main learning context. This highlights the potential of experiential and place-based learning to enrich geometry instruction (Waite, 2010). Encouraging PMTs to explore such strategies could diversify teaching methods and increase pupil engagement.

Regarding assessment, reliance on self-assessment techniques such as exit cards indicates an awareness of formative assessment strategies. However, the limited range of assessment approaches suggests an area for growth. Research (e.g., Black & Wiliam, 2018) emphasises the importance of diverse assessment tools, including peer assessment and structured rubrics, to evaluate reasoning and proof effectively. Expanding PMTs' repertoire of assessment methods could improve their ability to measure pupils' progress in reasoning and proof.

The minimal use of digital tools (only one group referenced GeoGebra) highlights a gap in leveraging technology to support mathematical reasoning. This may be because, in many countries (such as Slovakia, the Czech Republic, and Italy), teachers must reserve a special room for pupils to use digital tools. Consequently, the teacher often uses digital tools for demonstration purposes during mathematics lessons, which is more feasible in the countries mentioned. This contrasts with research advocating for technology as a critical tool in visualising and exploring mathematical concepts (Drijvers et al., 2024; Hoyles & Noss, 2019). Dynamic geometry software, for example, can enable pupils to manipulate and investigate geometric relationships interactively, supporting both experimental and deductive reasoning. Increasing PMTs' confidence and proficiency in using such tools could enhance the effectiveness of their teaching.

More generally, evidence from the international PMTs Meeting showed the transformative potential of international collaboration in teacher education. Hearing diverse voices and encountering varied educational perspectives proved an eye-opening experience for PMTs. Engaging in collaborative activities within mixed international groups fostered a deeper appreciation for the multifaceted nature of R&P and its pedagogical approaches. This aligns with previous studies (e.g., Jaworski & Huang, 2014), which underscore the value of international cooperation in broadening PMTs' and teachers' understandings. Also, working in a heterogeneous community of PMTs can support the development of communities of practice (Forman, 2020). As Clarke and Hollingsworth (2002) discussed in their interconnected model of teacher professional growth, such experiences enable participants to extend their reflective practices by integrating knowledge from multiple contexts. Moreover, Coles and Helliwell (2023) emphasise the role of internationalised teacher education in preparing teachers to address global challenges in mathematics education through culturally responsive task design.

Overall, the findings reveal both strengths and areas for improvement in PMTs' lesson planning. Their emphasis on pupil-centred learning, collaboration, and teacher facilitation is commendable. However, gaps in addressing misconceptions, integrating technology, and explicitly embedding reasoning into LPs point to opportunities for enhancing teacher preparation programmes. More recent research reinforces these conclusions. Weingarden and Buchbinder (2023) analysed how pre-service teachers adapted LPs to incorporate R&P, identifying several dimensions of modification, including logic-based and reasoning-oriented changes. Similarly, Buchbinder et al. (2023) demonstrated that a focused module on proving significantly enhanced PMTs' mathematical knowledge for teaching proof and their ability to justify solutions in lesson design. These results highlight the importance of targeted interventions to deepen PMTs' understanding of R&P and to foster its integration across diverse mathematical topics and contexts.

6.3 The influence of the Intervention design on PMTs (RQ3)

Our findings on **reading** revealed that several pedagogical tools introduced during the Intervention were well received by PMTs, particularly those supporting lesson planning. Van den Akker's (2013) "spider web" model was positively regarded across participating countries, aligning with research by Boon et al. (2022) that shows that clear frameworks are likely to help teachers develop cohesive and effective LPs. Equally, Remillard (2005) highlights how such tools can help teachers organise lesson components cohesively. However, some Norwegian PMTs reported difficulties with the terminology, which differed from what they used in their home institution. This points to the importance of providing materials in clear, accessible language to accommodate diverse linguistic backgrounds, as noted by Chun and Abdullah (2022). Hence, while some PMTs found the tool unfamiliar or overwhelming, their positive responses align with Perry and Lewis (2009), who note that comprehensive planning tools can be challenging for teachers without prior experience but can become more effective with further training.

Ellis et al.'s (2019) work on teacher moves, focusing on strategies that guide students' mathematical reasoning, was also highly praised. Many PMTs were surprised at the impact these moves could have on students' understanding, supporting research by van Es and Sherin (2008), highlighting the role of questioning and scaffolding in fostering mathematical thinking. Some PMTs expressed a desire for more time to explore teacher moves, reflecting the findings of Anthony and Walshaw (2009), who argue that sustained professional development is crucial for deepening teachers' understanding of these strategies.

In contrast, due to its complex nature and language, the text on socio-mathematical norms was challenging for some PMTs, particularly those from Czechia and Italy. This suggests that theoretical readings, while valuable, may be difficult to grasp without sufficient support or contextualisation. Yenmez and Celik (2021) emphasise the need for grounding theoretical concepts in concrete examples and discussions to help PMTs connect theory to practice. Introducing socio-mathematical norms through more intuitive, interactive activities before assigning complex readings would likely improve PMTs' understanding.

The text on modes of reasoning by Silverman and Even (2015) posed significant challenges for PMTs in our Intervention, many of whom struggled to differentiate between modes of reasoning without explicit discussion. This resonates with research by Stylianides (2019), highlighting the difficulties teachers face in understanding and applying abstract theoretical concepts without clear explanations and practical examples.

PMTs' reflections on **task redesign** highlight the transformative potential of task redesign in mathematics education. Many expressed surprise at how small changes could significantly impact students' learning, echoing Stein et al.'s (2009) emphasis on tasks with high cognitive demand that deeply engage students with mathematical concepts. This supports the PMTs' realisation that even minor adjustments can make tasks more intellectually engaging and foster R&P skills.

The PMTs also recognised that many tasks, even those initially seen as low-cognitive demand, could be restructured for higher-level thinking, aligning with Smith and Stein (2018), who stress the importance of rich tasks that encourage mathematical reasoning and proof. The reflections suggest that task redesign is not just about increasing difficulty but enhancing tasks' potential for deeper mathematical exploration.

PMTs' reflections on task redesign to focus on reasoning and proof resonate with Stylianides and Stylianides (2009), who emphasise the importance of integrating reasoning and proof throughout the mathematics curriculum. The PMTs' realisation that many tasks could be restructured to focus on R&P aligns with broader literature advocating for R&P as fundamental components of mathematical learning.

Collaborative lesson planning emerged as a key aspect of the Intervention, with many PMTs valuing the opportunity to work in mixed groups. The presence of more experienced peers in some groups helped PMTs understand potential student difficulties, aligning with Borko et al. (2010), who found that collaboration with experienced teachers enriches the planning process by offering practical insights into classroom challenges. Research by Williams et al. (2023) highlights how cross-cultural learning environments encourage teacher candidates to negotiate and integrate diverse teaching strategies, fostering innovative approaches to classroom practice. By working collaboratively in international teams, participants developed their design capacity and gained a deeper appreciation of the global dimensions of mathematics education. Xu and Connelly (2022) provide further evidence, showing how reciprocal learning frameworks enable teachers to draw on the strengths of different educational systems. Additionally, the work of Vale and Barbosa (2024) or Zaslavsky et al. (2011) on collaborative task design emphasises how international learning opportunities can lead to the creation of more sophisticated and culturally adaptable teaching strategies.

However, challenges arose during group work, such as time constraints and differing opinions on approaching the lesson. These difficulties are consistent with Grossman et al.'s (2001) research on professional collaboration, which identifies challenges in reconciling varying views and working styles.

In the Final Assignments, most PMTs focused on preparing their LPs rather than **reflecting on other PMTs' plans** or the enacted lesson. Reflection appeared to be seen as a routine part of the learning process, not a standout feature of the Intervention. This is likely because reflecting on peers' work is common for PMTs in all countries. The one-time reflection on the enacted lesson may have contributed to its omission in most PMTs' responses. Schön (1983) and Van Es and Sherin (2008) argue that reflection is essential for professional development, but its depth depends on context and frequency.

A few exceptions arose, with some PMTs appreciating the chance to reflect on their planning decisions, highlighting the role of reflection in improving teaching practices, as emphasised by Korthagen (2004). One PMT also found value in reflecting on the enacted lesson through video discussions, supporting Borko et al.'s (2008) notion that such reflection helps refine instructional decisions.

Reflection, particularly on enacted lessons, emerged as a key theme. Reflections provided deeper insights by allowing PMTs to compare practices across countries, fostering a comprehensive professional development cycle. Norwegian PMTs particularly valued the reflection on the enacted lesson in Session 4 as a fitting conclusion to the Intervention, embedding the programme's features and enhancing their teaching understanding. Guskey (2002) underscores the importance of reflecting on enacted lessons, especially through peer discussions, for refining instructional decisions.

Feedback, especially from peers and instructors, was frequently mentioned as helping PMTs revise their LPs. Hattie and Timperley (2007) stress the importance of feedback in teacher development, providing actionable insights that foster growth. Many PMTs appreciated both positive and negative feedback, aligning with Shute's (2008) assertion that constructive feedback can significantly improve learning outcomes and professional growth.

Participants highlighted the benefits of **working in mixed, international groups**, which exposed them to diverse pedagogical approaches and teaching contexts. These interactions broadened their understanding, enriched lesson planning, and fostered idea-sharing, critique, and refinement. This aligns with Borko et al. (2010), who stress that collaboration expands teaching strategies and enhances the ability to address student challenges. Similarly, Stein et al. (2008) and Eurydice (2020) highlight that cross-cultural collaboration deepens instructional dialogue and enhances reflective practices.

Although initial challenges such as unfamiliarity and language barriers arose, the collaborative process ultimately improved PMTs' understanding of effective lesson structure.

Integrating R&P into lesson planning was another focus. International collaboration revealed how tasks could be redesigned to embed R&P, fostering mathematical thinking. This aligns with Stylianides and Stylianides (2009), who advocate for embedding R&P across the curriculum. Collaborative planning sessions and cross-cultural feedback refined PMTs' approaches, promoting higher cognitive engagement and enriching teaching practices.

7

CONCLUSIONS, LIMITATIONS, AND IMPLICATIONS FOR PRACTICE

In this chapter, we first discuss the centrality of reasoning and proving (R&P) in mathematics education, examining how different student perceptions and international contexts influence R&P practices. We then consider the role of pre-service mathematics teachers' (PMTs') engagement with R&P in lesson and task design, highlighting how meaningful training opportunities can foster exploration, justification, and argumentation in the classroom. Next, we explore the impact of international and collaborative learning experiences on PMTs' professional growth and design capacity. Finally, we reflect on the theoretical and practical contributions of the MaTeK project's Intervention, including its iterative cycles, sustainability, and implications for preparing teachers to navigate diverse educational contexts.

First, our study results underscore the centrality of R&P as a cornerstone of effective mathematics education. In addition, they provide explanations and examples of different student perceptions of R&P and R&P practices in different national contexts. Teachers' ability to incorporate R&P into their lesson planning is crucial in promoting students' deeper understanding of mathematics. When PMTs have meaningful opportunities to engage with R&P during their training, they are more likely to create tasks that encourage exploration, justification, and argumentation. Importantly, how teachers perceive the nature of proof, whether as a rigid logical structure or as a more dynamic, investigative process, can significantly influence how they bring R&P into their classrooms. This highlights the importance of early support in developing a clear and flexible understanding of R&P, which is in line with the goals of the Intervention, which focused on strengthening lesson and task design that foreground reasoning and evidence-based thinking.

DOI: 10.4324/9781003533870-7

Secondly, the research highlights the potential of international learning experiences in enhancing the design capacity of PMTs. It suggests ways of working with PMTs nationally and internationally. Engaging with R&P in mathematics teaching across diverse educational contexts exposed participants to various pedagogical approaches and cultural perspectives, enriching their understanding and practice. International collaborations provide unique opportunities for teacher candidates to learn from diverse approaches to mathematical tasks and instruction. Such experiences support teachers in deepening their reflective practice by connecting insights gained across different contexts. They help participants understand and integrate various forms of knowledge, contributing to their ongoing professional growth. Additionally, engaging with diverse educational settings encourages teachers to consider global perspectives and respond to cultural differences in their task design, better equipping them to address contemporary challenges in mathematics education.

Collaborative learning across borders emerged as a key enabler of professional growth for PMTs. The cross-national nature of this Intervention fostered meaningful interactions, allowing participants to reflect on their teaching practices and adapt them to new cultural and educational settings. Cross-cultural learning environments encourage teacher candidates to engage with various teaching approaches, leading to more innovative and adaptable classroom practices. Collaborative work in international teams enhanced participants' ability to design effective tasks and lessons and deepened their understanding of the global nature of mathematics education. Reciprocal learning across different systems allowed them to draw on a broad range of strengths, while joint task design fostered the development of more refined and culturally responsive teaching strategies.

Third, in methodological as well as practical terms, the Intervention of the MaTeK project provides a tested and validated prototype of an intervention in terms of R&P. Leaning on students' evaluations, the Intervention can be said to have been successful, in particular in raising PMTs' lesson design capacities (interpreted as part of pedagogical content knowledge). Regarding sustainability, the second and third cycles of the Intervention designed within the MaTeK project have already been realised and play a critical role in maintaining the relevance of PMTs' preparation in the ever-evolving field of mathematics education. These iterative cycles allowed for the refinement of the Intervention based on feedback and data from previous implementations, ensuring that it remains responsive to the needs of both teachers and students. By incorporating updated pedagogical strategies and addressing emerging challenges in R&P practices, these cycles help PMTs stay current with innovative teaching methods and align with international trends in mathematics education. Furthermore, the repeated cycles facilitate the institutionalisation of the Intervention within teacher education programmes, creating a framework for continuous professional development that extends beyond the duration of the MaTeK project. This approach not only ensures the long-term impact of the project but also supports the ongoing preparation of PMTs to meet the demands of diverse and dynamic educational contexts.

Fourth, it can be claimed that, theoretically, the concept of "design capacity" has been enlarged. Through the Intervention, design capacity emerged not only as the ability to create tasks and lesson plans (LPs) but also as a dynamic competence that integrates content knowledge, pedagogical reasoning, and sensitivity to cultural and contextual factors. PMTs demonstrated that effective design extends beyond technical planning; it involves critical reflection, adaptation, and the ability to anticipate and respond to students' thinking. The collaborative and cross-cultural dimensions of the project further revealed how design capacity can be nurtured through dialogue, co-construction of meaning, and exposure to diverse educational traditions. This expanded understanding positions design capacity as a key element of teacher professionalism and a foundational component of preparing teachers to meet the complex demands of contemporary mathematics classrooms.

In conclusion, the findings from the study underscore the transformative potential of international collaboration in teacher education, particularly in lesson planning and feedback. By working together across countries, PMTs could enhance their pedagogical content knowledge, refine their LPs, and engage in meaningful reflections on their teaching practices. This collaborative process improved their understanding of mathematics education and highlighted the importance of R&P in fostering mathematical thinking. The study illustrates that international cooperation, combined with structured lesson planning and constructive feedback, provides a powerful mechanism for enhancing teacher professional development, particularly in lesson design capacity.

7.1 Implications for practice

This part outlines the practical implications of the study's findings for mathematics education. It highlights how teachers, curriculum developers, or policymakers may apply the results to enhance teaching and learning in mathematics teacher education, particularly regarding supporting PMTs in their lesson design. Drawing connections between the research outcomes and real-world classroom settings, the chapter emphasises actionable strategies and potential shifts in instructional practice. In addition, it provides a critical reflection on the study's methodological and contextual limitations that may affect the interpretation or generalisability of the results. Acknowledging these limitations ensures transparency and guides future research to build on the current study's contributions.

PRACTICE 1: PROVISION OF PRACTICAL PEDAGOGICAL TOOLS FOR PMTS

A central outcome of the Intervention was that PMTs highly valued the opportunity to gain practical pedagogical tools that could be directly applied in their classrooms. Specifically, they appreciated the "teacher moves" (Ellis et al., 2019) demonstrated during the Intervention, which offered concrete strategies for fostering student reasoning and supporting the development of R&P. These tools, such as questioning techniques, prompts for justifying solutions, and strategies for scaffolding student thinking, were seen as essential for creating a classroom environment conducive to R&P. By observing teacher moves in action, PMTs were able to visualise how different pedagogical strategies could be employed to help students engage with and deepen their understanding of R&P.

Moreover, demonstrating various ways of reasoning on tasks from different educational contexts, specifically from different countries, proved invaluable. This cross-national approach provided PMTs with a broader repertoire of examples and strategies that they could adapt to their own teaching contexts. The diversity of educational settings highlighted how reasoning could be approached in different ways depending on the cultural and pedagogical norms, allowing PMTs to reflect on how to tailor their teaching approaches to meet the needs of their own students. This aligns with previous research that underscores the importance of providing teachers with adaptable, context-sensitive tools to enhance their instructional practice (Bergqvist & Lithner, 2012; Boesen et al., 2010, 2014).

PRACTICE 2: CONNECTION OF THEORY AND PRACTICE

The second key principle emerging from our study is the importance of connecting theory with practice. For PMTs, seeing their LPs "come to life" during the Intervention was crucial in bridging the gap between theoretical knowledge and practical application. Enacting selected LPs was motivating from the very beginning and fostered a sense of healthy competition among groups, as each aimed to create the most effective LP. An additional benefit was the opportunity to work with real pupils and receive feedback from the school teacher during enactment, which provided authentic insights into the lesson's effectiveness.

This connection between theory and practice aligns with literature on effective teacher education, which highlights the central role of reflection and peer collaboration in improving teaching practices (Dewey, 1933; Schön, 1983). Through the enactment process, PMTs were able to critically observe how pupils responded to tasks, identify which aspects of their planning worked well, and recognise areas needing adjustment. This hands-on experience encouraged continuous reflection on instructional decisions and helped PMTs link abstract theoretical ideas about R&P to concrete classroom interactions. Such reflective engagement strengthens teachers' pedagogical content knowledge and enhances their capacity to adapt teaching strategies to real classroom dynamics (Forsler et al., 2024; Hußner et al., 2023).

PRACTICE 3: IMPORTANCE OF INTERNATIONAL MEETINGS AND DISCUSSIONS

The cross-national variability in perspectives and approaches proved instrumental in enhancing the PMTs' conceptualisation of R&P. By engaging in mixed international groups, PMTs were exposed to a wide array of teaching practices, pedagogical strategies, and cultural contexts, all of which enriched their understanding of R&P and lesson planning. These interactions facilitated a deeper appreciation for the diverse ways R&P can be integrated into the mathematics curriculum, highlighting the importance of context, language, and the varying levels of cognitive demand required to teach R&P effectively. The eye-opening nature of these international exchanges points to the transformative potential of collaborative professional development. By engaging in cross-national dialogue, PMTs not only broaden their conceptualisation of R&P but also gain valuable insights into teaching R&P in culturally responsive and pedagogically effective ways. This underscores the importance of international cooperation in advancing mathematics education and improving the teaching of complex topics such as R&P.

These principles are fundamental to the development of PMTs. The Intervention demonstrated that equipping PMTs with adaptable, context-sensitive tools and creating opportunities for reflective practice through peer and educator discussions are key to enhancing their understanding and teaching of R&P. These principles also support the notion that effective teacher education must balance theoretical knowledge with practical application, ensuring teachers are prepared to engage students in meaningful and effective learning experiences.

7.2 Limitations of the study

The brevity of our Intervention may also have limited its impact. Spanning just four sessions, the Intervention provided limited time for PMTs to fully internalise and apply the principles of R&P in their LPs. Comparatively, Buchbinder and McCrone's (2023) year-long Intervention demonstrated more pronounced outcomes, suggesting that longer, more sustained interventions may yield deeper transformations. In our case, significant changes in PMTs' approaches to R&P emerged during the international meeting at the end of the Intervention, highlighting the potential benefits of extending the Intervention period and integrating diverse perspectives through collaborative international work.

Another limitation was that we could not follow the Intervention participants after the Intervention, for example, in their school placement, to see what they learned, what LPs they designed, how they enacted the lessons, and how they incorporated R&P.

Extending the duration and providing sustained support for PMTs could be beneficial to maximise the impact of such interventions. Moreover, fostering opportunities for international collaboration, as highlighted by the positive outcomes of the international meeting, could further enhance participants' capacity to integrate R&P into lesson planning. These findings underscore the importance of aligning Intervention design with the background and needs of participants, as well as incorporating research-backed strategies to support meaningful changes in PMTs' pedagogical practices (Stylianides & Stylianides, 2017, 2022; Zaslavsky, 2019).

8

COMMENTARY

[Im]Proving classroom mathematical reasoning through a lesson-planning intervention

Kenneth Ruthven

This book reports on the MaTeK (Enhancement of Research Excellence in Mathematics Teacher Knowledge) project. That project was conducted by didactician researchers from five universities which prepare future secondary mathematics teachers – each university in a different European country. It provides a fine example of the spirit of ERME (the European Society for Research in Mathematics Education), the organisation associated with the series in which this book appears. That spirit is one of communication between different countries, of cooperation between different approaches, and of collaboration between researchers differing in experience.

The focus of the project was on the joint design and implementation of an intervention aimed at developing a particular aspect of the professional knowledge and expertise of pre-service mathematics teachers – relating to the planning of lessons in which mathematical reasoning and proving play a central part. The aim was to provide participants with tools to [im]prove classroom mathematical reasoning – improve in the sense of enriching such reasoning through informed lesson planning; prove in the sense of demonstrating the character of such reasoning by means of the framework of constructs provided by the intervention.

This commentary chapter will first discuss the conceptualisation of the project, the instrumentation of the intervention, and the context(s) in which these were undertaken. It will then offer an evaluation of the intervention, focusing on two questions: Did the pre-service teachers participating in the project appropriate the apparatus of lesson planning promoted by the intervention? Did participants develop their understanding of pedagogical aspects of reasoning and proving?

DOI: 10.4324/9781003533870-8

8.1 Conceptualisation of the project

The Introductory chapter sets out the ideas motivating the project. The foundational notion is that reasoning and proving are central to learning mathematics with understanding. In the ensuing Theoretical chapter, mathematical reasoning emerges as an elastic term which has attracted many conceptualisations. The most expansive of these encompasses multiple facets: engaging in deep thinking; communicating about such thinking with others; devising strategies to tackle unfamiliar problems; linking different mathematical ideas; forming conjectures and testing them; and using logical arguments to validate conjectures. A much narrower, if more sharply defined, characterisation separates out this last component (or, more liberally, the last two components) from the others, identifying mathematical reasoning with the practice of proving. In particular, proof can be pedagogically attractive as a means of deepening mathematical knowledge and understanding when it not only conveys methods and strategies but explains why results are true.

The other main idea motivating the project is that teaching and learning mathematical reasoning and proving are widely acknowledged to be challenging. Indeed, the suggestion is that mathematics teachers themselves often lack knowledge and experience relevant to preparing and teaching lessons in which mathematical reasoning and proving play a central part. The Introductory chapter argues that even where teachers do have some such knowledge and experience, it may be confined to what they find in school textbooks. Certainly, such considerations would imply that there is a particular need to fill such gaps in the knowledge and experience of pre-service mathematics teachers whose "apprenticeship of observation" as students is likely to have provided them with only restricted models of reasoning and proving in the school classroom. The chapter implies that this is not so much an individual weakness as an institutional one of the professional field, arguing that there is limited didactical knowledge on how teachers can successfully create learning environments that support student engagement with reasoning and proving. Consequently, the project proposes to advance theoretical and practical knowledge on preparing teachers to plan and enact this type of mathematics lesson.

The Theoretical chapter introduces the key processes guiding the project's plan of action as those of lesson planning and task (re)design, conducted in a collaborative and reflective manner, and its key concept as that of "lesson design capacity". In particular, the suggestion is that lesson planning and task (re)design can be educative processes, aimed not just at the immediate effectiveness of teaching but at the longer-term professional learning of the teachers undertaking them. The argument is that the "didactical fluency" of the "teacher-as-designer", namely their "lesson design capacity", can be developed through such a process, particularly if undertaken in collaboration, and with opportunities for joint reflection. Consequently, the collectively designed project intervention centres on a sequence of sessions intended to support pre-service mathematics teachers in learning to plan lessons in which mathematical reasoning and proving play a central part. The Methodology chapter sets out how this sequence is organised around an iterative process in which the pre-service teachers plan, review, and revise lessons, drawing on a range of resources.

8.2 Instrumentation of the intervention

These resources are of two main types. First, pre-service teachers are invited to select existing curriculum resources which provide tasks that they consider relevant to reasoning and proving, and to then adopt or adapt them through a process of lesson design. Second, the project has selected certain pedagogical devices intended to serve as tools for the participants in designing and implementing lessons. These devices focus on (1) components of lesson planning, (2) interactional moves in teaching mathematics, (3) modes of mathematical reasoning, and (4) classroom socio-mathematical norms. The intention is that the set of constructs provided by each of these pedagogical devices should become a functional tool for the pre-service mathematics teachers through a staged process in which they first read texts about particular devices, and then gain experience of putting them to use in the course of iteratively planning, reviewing, and revising lessons. This learning process is supported by class sessions in which each device receives particular attention, sometimes aided by a protocol or template prepared by the project team, as well as observations and video-recordings of enacted lessons. These texts, aids, experiences, and records are intended to serve, then, as a system of resources through which use of the various pedagogical devices is developed and coordinated to form an overarching instrument of lesson design. The intention is that such an instrument should support project participants in creating learning environments that support student engagement with reasoning and proving.

Indeed, in terms of the enlarged construct of "resource" outlined in the Theoretical chapter, further entities would also be considered to be resources contributing to the learning process – notably the interactions during sessions, particularly feedback from others. These have potential to stimulate reflection, sometimes itself recorded in writing to preserve what might otherwise be ephemeral. Together, then, this whole system of resources is intended to provide project participants with a framework both for the immediate design and implementation of lessons, and for longer-term professional learning.

8.3 Context(s) of the project

As already noted, the project involved collaboration between five university departments, each in a different European country. The Background chapter offers a description of each context – and a comparison of them. The focus is on the national programmes of school mathematics and the university programmes of teacher education. Particular attention is given to those aspects relevant to planning lessons related to mathematical reasoning and proof.

This chapter reports that, in all the countries, the school mathematics curriculum gives some attention to reasoning, logical thinking, and argumentation, but generally treats them as strands within broader competencies such as problem-solving and communication. However, the depth and explicitness of such integration, as well as the emphasis on different aspects of reasoning and proof, vary significantly. At one extreme, the Norwegian curriculum integrates reasoning through broad process-oriented goals while making minimal reference to proof. By contrast, the Slovakian curriculum incorporates a dedicated section on logic, reasoning and proof. In the curricula of other countries – Czechia and Italy – reasoning and proving are particularly associated with certain mathematical domains, notably logic and geometry.

This chapter also highlights some notable differences between the teacher education programmes that the pre-service teachers in the five institutions had been following prior to participating in the project. One fundamental difference was in the character of the mathematically related programme components – a difference with implications not just for the mathematical knowledge gained by the pre-service teachers but for the professional identity formed by them. Most strikingly, project participants in the Norwegian university – engaging in programmes of teacher education emphasising schooling of the whole person – had studied no conventional courses in mathematics as such, but rather courses specifically in mathematics education. In all the other universities, conventional mathematics courses had constituted a substantial portion of the curriculum. Indeed, project participants in the Italian and Slovakian universities – engaging in programmes of teacher education emphasising academic subject specialism – had followed no mathematics education courses before the intervention.

The evidence suggests that this had provided contrasting experiences of reasoning and proving. In the mathematics education courses in the Norwegian institution, pre-service teachers encountered "generalisations, conjecturing, argumentation, justifications, and proving related to school mathematics topics",[1] while "[t]he proving methods utilised included mainly visual representation proofs and generic examples". However, these students did "experience [] formal proofs and formal logic to a limited extent within a course focusing on school geometry, where they learned about Euclidean geometry and the Euclidean axiomatic system". In Slovakia, by contrast, the pre-service mathematics teachers had followed a range of conventional advanced mathematics courses with a strong emphasis on formal proof, so that they "encountered various types of reasoning and proof (e.g., direct, indirect, proof by contradiction)", where "[t]he aim was to teach them not only to understand proofs but also to be able to create them independently".

Accordingly, the Methodology chapter reports that devising a common Intervention – one judged capable of being implemented successfully in all five countries – was a demanding process. The project coped with these demands in two ways. On the one hand, the planned Intervention accommodated differences through the scope it offered pre-service teachers to adapt tasks from existing curriculum resources, which could originate in their own context and reflect its affinities, rather than specifying particular tasks. On the other hand, the Intervention distanced itself from any particular context through its adoption of pedagogical devices and discourses drawn from an international research literature.

8.4 Evaluation of the intervention

How successful was the resulting Intervention? In my view, given the emphases of the project, there are two particularly important evaluative questions. First, did the pre-service teachers participating in the project appropriate the apparatus of lesson planning promoted by the Intervention? Second, more broadly, did participants develop their understanding of pedagogical aspects of reasoning and proving?

8.4.1 *Did participants appropriate the apparatus of lesson planning promoted by the intervention?*

One source of evidence about the extent of participants' appropriation of the apparatus of lesson planning can be found in the analysis of their experience of the Intervention, reported in the Findings chapter. This indicates that the four pedagogical devices were received with very different levels of enthusiasm. Response to the "spider web" device for lesson planning was generally very positive, although there was some resistance from participants who were already conversant with alternative models. While widely appreciated, the "teacher moves" device was found challenging to access by some participants. Many participants did not find the constructs of the "reasoning modes" device easy to distinguish in practice, while others were uncertain about how to use the resulting insights in practice. Finally, participants made little reference to the "socio-mathematical norms" device, and it appears that few found it accessible. In terms of participants' opinions, then, it seems that only one of the four pedagogical devices promoted by the Intervention was strongly appropriated by project participants, and one other more weakly so.

It is possible to triangulate what participants said against what they proposed to do by examining the accounts and analyses of lesson plans reported in the Findings chapter. These reports relate to the version of lesson plans created after all the pedagogical devices had been discussed in sessions. Looking at the degree to which each device features in these reports on lesson plans is a plausible indicator of the extent to which they had become functional tools for the participants.

While there is only one explicit reference to the "spider web" there, it seems that analysis of the lesson plans by didactician researchers was guided in part by the components of this lesson-planning tool, suggesting that these components be used as indicators of the degree to which this device had been appropriated by participants. Here, the evidence is that all the lesson plans were attentive to the components of "Aims and objectives", "Content", "Learning activities", and "Teacher and student role". The most striking absence in the lesson plans was attention to the "Assessment" component – "there were few mentions and details about assessment in most of the LPs, except for the Turkish LP where a distinction was made between 'short-term' (exit cards where pupils could describe their difficulties) and homework/longer-term". A further insight from the analysis of changes in the lesson plans between versions – changes either of substance or in making matters explicit – is that the majority related to the components "Learning activities" and "Aims and objectives", suggesting that these were the components of lesson planning most prominent in the participants' evolving thinking. All in all, then, the evidence points to a relatively strong appropriation across participants of the "spider web" tool for lesson planning.

Evidence of uptake of the "teacher moves" device was more varied, ranging from absence to explicit presence: "no teacher moves were listed in the [Czech] LP; the [Italian] lesson plan ... referred to selected questions for pupils and expected answers [but] did not always explicitly outline teacher moves"; the Slovakian plan "explicitly listed several teacher moves (Ellis et al., 2019)"; the Turkish plan "explicitly outlined many teacher moves"; in the Norwegian plan, "teacher moves were identified using Ellis et al.'s research (2019)".

There is no explicit reference to "reasoning modes" in any of the national lesson reports. Just one hints at some influence of the constructs from this device: "[the Czech plan] anticipated a few ways of pupils' reasoning (e.g., appeal to authority, providing a counterexample)". The cross-national comparison of lesson plans examines reasoning without making any reference to the detailed constructs listed in the "reasoning modes" text. Finally, while there was passing reference to norms in two of the reports on lesson plans, there was no evidence of explicit use of the constructs of the "socio-mathematical norms" device. Equally, the cross-national comparison of lesson plans reports that "[t]he text on socio-mathematical norms was rarely mentioned in the data".

This triangulation supports the conclusion that only one of the four pedagogical devices was strongly appropriated by participants to become a functional tool, and one other more weakly so, with the two further devices not taken up to any significant degree. The Discussion chapter is surely correct in suggesting that "[t]he brevity of our Intervention may also have limited its impact... [It] provided limited time for PMTs to fully internalise and apply". A further explanation offered is that the texts explaining the latter two devices were not sufficiently accessible to participants. Beyond these observations, one might point to the considerable challenge not just of coming to terms with four unfamiliar devices at much the same time, but also of forming an integrated operational understanding of them as a combined system. In the language of French didactique, there is a complex "instrumental genesis" to be analysed and orchestrated, covering activation of each of the four distinct pedagogical devices and their transformation into a coordinated unitary instrument. In my view, this is a fundamental issue which any future development of the Intervention ought to take into account.

8.4.2 Did participants develop their understanding of pedagogical aspects of reasoning and proving?

The choice of the "teacher moves", "reasoning modes", and "socio-mathematical norms" devices for inclusion in the apparatus for the Intervention reflects its particular concern with pedagogical support for classroom mathematical reasoning. Even if the framework of constructs for analysing classroom mathematical reasoning that these devices provided was not appropriated by project participants to any great degree, the Intervention as a whole may have influenced their understanding of pedagogical aspects of reasoning and proving. Here again, it is possible to triangulate what participants said and what they proposed to do, on this occasion by examining the reports of the interviews conducted with groups of participants (what they said) and the accounts and analyses of lesson plans (what they proposed to do).

The cross-national comparison of the evolution of lesson plans from their initial versions to the final adaptations, as revised after enactment, suggests that the iterative lesson-planning process at the heart of the Intervention had little influence on participants' treatment of reasoning and proving: "while [participants] made various modifications to instructional strategies, assessment methods, and classroom activities based on their enactment experiences, the way reasoning was incorporated remained largely unchanged". In part, this may be because "despite the explicit assignment of designing and enacting a lesson that fosters R&P, its presence was often implicit rather than explicitly structured within the plans".

The first part of the interview analysis explored participants' notions of reasoning and proof.[2] As noted earlier, in some countries mathematical reasoning and proof are particularly associated with particular mathematical domains, notably logic and geometry, where they are often treated in a rather formal manner. The Findings and Discussion chapters highlight a particularly significant transformative impact of the Intervention on these aspects of participants' beliefs about reasoning and proof, comparing responses before and after the Intervention:

> Participants, especially those from Slovakia (SK) and Italy (IT), experienced a significant shift in their understanding of R&P. Initially seen as a distinct and isolated topic ... PMTs began to appreciate the presence of R&P throughout the curriculum and its informal and exploratory dimensions, moving beyond an exclusive focus on formal proof.

The next part of the analysis explored interviewees' beliefs about the role of reasoning and proof in the teaching-learning process. The Findings chapter reports that the same five roles emerged in both pre- and post-interviews, but that in the latter "the PMTs spoke about them with more conviction". That apart, there is little evidence of any substantial change in participants' understandings.

The final part of the analysis explored interviewees' concerns about teaching reasoning and proof. Here, a tension emerges between the pedagogical role in which reasoning and proof is seen as "lead[ing] to pupils' engagement with mathematics" and practical concern about "pupil difficulties" in understanding reasoning and proof. There is evidence of some change in this respect. In the pre-interviews, there seems to have been general agreement that reasoning and proof "was primarily designed for pupils considered intellectually advanced or exceptionally talented". While, after the Intervention, some participants remained convinced that there were pupils for whom reasoning and proof was too difficult, others had shifted to a view that "certain types of arguments were suitable for different levels of ability". Equally, after the Intervention, participants "increasingly acknowledged the teacher's responsibility in cultivating a reasoning-oriented classroom through questioning and interaction, rather than expecting such norms to develop independently".

In summary, while lesson plans provided little evidence of the iterative planning process having an impact on participants' pedagogical understanding of reasoning and proving, interview responses pointed to the Intervention as a whole as influencing aspects of such understanding. Significant changes were a broadening of conceptions of reasoning and proving, greater conviction about the pedagogical roles of reasoning and proving, some shift towards a more inclusive position on the accessibility of reasoning and proving to a wider range of pupils, and increasing recognition of the interactive role of the teacher in supporting reasoning and proving activity in the classroom.

Notes

1 All quotations are taken from the versions of the chapters that I received for commentary in July 2025.
2 "Reasoning and proof" is the combined phrase consistently used in the interview protocol. Because of this, I have used the term reasoning-and-proof as a singular entity in this section.

9

COMMENTARY*

Pre-service mathematics teachers as lesson designers

Luc Trouche

First of all, I would like to thank Mária Slavíčková for inviting me to write a short commentary on this book. It is an honour for me to thus contribute to the European research project MaTeK, which has brought together research teams and mathematics teacher trainers from five countries (the Czech Republic, Italy, Norway, Slovakia, and Türkiye).

Chapter 1 provides a clear introduction to the project, which builds on a strong theoretical foundation and a carefully designed methodology. Initially developed in each country, the project was enriched through the international exchange of ideas between researchers and practitioners.

What makes MaTeK particularly interesting is its focus on two central and challenging issues in mathematics education: the preparation of new teachers and the integration of reasoning and proof (R&P) at the heart of teaching. These issues have long been essential, but they are becoming even more critical today due to two converging factors: the shortage of mathematics teachers and the emergence of artificial intelligence, which increasingly influences how we approach mathematical reasoning.

This commentary adopts two complementary perspectives: reflecting on the MaTeK project in light of my own research experience and considering potential future directions for mathematics education in a rapidly changing technological and pedagogical landscape.

* Reflections on the book – Relevance of the project, key contributions, implications for the field, possible developments.

DOI: 10.4324/9781003533870-9

9.1 A theoretical framework combining different approaches

Chapter 2 presents the project's theoretical foundation. Considering the teacher as a designer is a particularly relevant choice, aligning naturally with the documentational approach to didactics (Trouche et al., 2020), which is central to several references cited in the project. Prospective mathematics teachers (PMTs) are not novice designers; even before entering the profession, they have already designed learning resources as students. This perspective resonates with the work of Assis et al. (2018), who study the transition from students' personal resource systems to those developed as teachers. Similarly, Zhu et al. (2025) adopt this approach to examine the professional growth of novice teachers, highlighting the need for a comprehensive framework to analyze both individual and collective dimensions of professional development.

I also appreciated the construction of a theoretical frame for studying R&P, particularly drawing on the framework proposed by Stylianides and Stylianides (2009), which integrates mathematical, psychological, and pedagogical perspectives. Balacheff (2010) further enriches this discussion, emphasizing an important epistemological point: "Despite the emphasis on the key role of proof in mathematics, it must be remembered that at stake is not truth but the validity of a statement within a well-defined theoretical context."

Finally, I found the framework proposed by Ellis et al. (2019), Teacher Moves for Supporting Student Reasoning (TMSSR), particularly compelling. By categorizing teacher moves into eliciting, responding, facilitating, and extending, it situates teacher evolution along a continuum, highlighting their impact on fostering student reasoning. The related concept of mathematics-didactical flexibility, only briefly mentioned in the book, deserves deeper exploration.

I also appreciated the construction of a theoretical frame for studying R&P, using in particular the approach of Stylianides and Stylianides (2009), encompassing mathematical, psychological, and pedagogical aspects. Balacheff (2010) also proposed a didactical perspective for "Bridging Knowing and Proving in Mathematics". He drew attention to an epistemological issue: "Despite the emphasis on the key role of proof in mathematics, it must be remembered that at stake is not truth but the validity of a statement within a well-defined theoretical context". I would like to emphasize this point because I noticed that, in Chapter 2, the word "truth" appears almost as often as the word "validity".

Finally, in this chapter, I appreciated the framework of Ellis et al. (2019), known as TMSSR. This notion of teacher moves allows teacher evolution to be situated into four categories, eliciting, responding, facilitating, and extending, on a continuum according to their potential for supporting student reasoning. The idea of mathematics-didactical flexibility for classroom teaching, just evoked in this chapter, appears as a fruitful notion worthy of further development.

9.2 The background of the study

Chapter 3 provides a comparative overview of teacher education programmes across the five countries. The contrasts are striking; for example, Italy requires 114 credits in mathematics compared to zero credits in Norway, while Norway devotes 60 credits to mathematics education compared to none in Italy. Similarly, practical teaching experience varies from ten days in Italy to 95 days in Norway. These structural differences inevitably shape PMTs' professional preparation.

It would also be useful to learn more about the role of technology, resources, and collaborative work within teacher education – all of which are essential for developing R&P competencies.

Comparing curricula from an R&P perspective reveals more similarities than differences. However, understanding how these curricula evolve over time is equally important. For example, Türkiye's curriculum introduces a stronger emphasis on R&P, while recent reforms in China encourage creativity and imagination in proof processes, contrasting with France's focus on explicit instruction (Trouche, 2025).

These trends align with broader international developments, notably the influence of the OECD (2023), which emphasizes student agency, computational thinking, modelling, and plausibility assessment – all central to R&P integration.

9.3 A well thought methodology

The methodology is carefully presented. The idea of initiating the project by establishing a common theoretical basis, based on reading a set of references, seems to me to be entirely relevant (and often underestimated in other projects). The alternation of design, reflection, and implementation phases (initially optional) is consistent with the theoretical framework that has been established. The implementation of lessons is monitored through video recordings. One could also imagine that, as with this type of system in China, all PMTs attend the lesson live in class to witness its implementation. Obviously, material conditions do not always make such live monitoring possible.

Finally, launching the project at the national level with a view to achieving results in an international context seems to be a truly productive agenda. A range of protagonists are involved in the project, including PMTs, of course, but also teacher educators, "critical friends", researchers, and the teachers in whose classrooms the implementation of the situations will take place. This diversity is very interesting and is often at the heart of professional development programmes (Karsenty et al., 2024). It would undoubtedly be useful to say a little more about the role of each protagonist.

The suggested readings are well chosen, combining theoretical and practical contributions (e.g., the "spider web" concept and Rationale for Teaching). One could also have included readings that issue the role of technology in the evidence process (e.g., Trouche, 2016), or readings that highlight the role and conditions of learning communities (e.g., Wenger et al., 2002), but the creation of a necessarily limited set of readings requires choices to be made (being aware that the presence or absence of certain readings will have consequences for the development of the project).

The programme is well structured, alternating between individual work, small group work, and then full group work (at the national level). It would be interesting to know a little more about how the groups are formed, the roles of each member, and the collaborative work tools (a platform, for example) that allow all protagonists at the national and international levels to interact throughout the project.

9.4 Results that inform the field of research

I appreciated the structure of Chapter 5 on results, which specifically addresses each research question.

The first research question concerns PMTs' conceptions of R&P before and after the intervention. The proposed classifications seem particularly productive to me:

- The role of R&P in mathematics: (a) justification of correctness; (b) deduction of new pieces of knowledge; and (c) the description of steps;
- The role of R&P in the teaching-learning process: (a) serving to justify the existence and truth of mathematical knowledge; (b) leading to pupils' engagement with mathematics; (c) leading to a deep understanding; (d) convincing pupils that something holds; and (e) helping to communicate mathematics;
- Concerns associated with teaching R&P: (a) curriculum; (b) time; (c) pupils' difficulties; (d) establishment of appropriate socio-mathematical norms in classes; and (e) PMTs' lack of prior experience and pedagogical content knowledge.

These categories, which emerged from the research project, should be very useful for future research. It would undoubtedly have been useful to highlight the conceptions before and after the project more clearly. However, to do so, and this could be the subject of further research, it would be necessary to focus particularly on a few individuals with contrasting profiles in order to highlight the relationships between views on R&P in mathematics, mathematics teaching, and concerns associated with R&P. Similarly, it would be interesting to relate these views to the curricula of each country.

The second research question is related to designing and implementing a lesson focusing on R&P. The structure of this section is complex: it first presents the third version of the lesson designed by each national group, then the lesson designed in mixed groups (bringing together PMTs from several countries), and finally the path from the first version to the third version, including a classroom implementation stage. However, some key themes clearly emerge.

For the third version of the lesson, common elements appear, such as the presence of digital or tangible manipulatives, the strong link with the curriculum, and the alternation of group work and whole-class discussions. There are also differences: R&P was addressed to varying degrees and through different conceptualizations; the role of the teacher varied (lesson pupil-centred except for the Italian case, which was teacher-centred). The authors note that this last difference could be due to the fact that the Italian team was inexperienced in lesson plan design. The notion of "path of reasoning" appears in the analysis; it would be interesting to explore this further.

For the lessons designed during the international session, the decision was made to set the subject area (geometry and measurement). Quite naturally, as this work comes at the end of the project, it feels like the PMTs' thinking is more advanced. Of course, the groups' work can no longer refer to a given curriculum; it would be interesting to know what tensions may have existed in each group between different national traditions, traditions that were clearly apparent in the first phase of the project. Another concept emerges in this analysis, that of "experimental demonstration", which would also be interesting to explore; is there a connection with Jon Borwein's experimental mathematics (2016)?

Finally, this section reviews the changes that occurred between successive lesson plans. The analysis is quite thorough, but we would like to know more about the implementation of the second version in classrooms. For example, we learn that

> the PMT who was the teacher and one of the LP's authors realised that Part 3 was not well connected with the rest of the lesson and with R&P. Some other PMTs noted during Session 4 that pupils saw the tasks in the worksheet as independent and that the teacher missed opportunities in the lesson to uncover the connections among them: this is a very interesting remark, evidencing different points of view according to the position of the PMT (in charge of implementing the lesson, or of its design, or simply attending the lesson, face to face, or via its video recording). We miss a global analysis of the "incidents

or surprises", occurring during the implementation, and leading to changes in the LP. A table like the one provided explaining the reasons for the changes from LP1 to LP2 could have been provided explaining the reasons for the changes from LP2 to LP3.

The third research question was related to how the PMT experienced the intervention.

I retain, as a first result, a new awareness of the importance of task design ("there is so much behind writing a task") and LP design ("using a few different words, a teacher can make a big difference in what pupils get from the interaction with the teacher").

I retain, as a second result, a new awareness of what is R&P and what should be R&P in mathematics teaching. I really liked the sentence that

> mathematics determines what is true. That in proving something, it's not about who expresses themselves better or who has a stronger opinion; we perceived the importance of reasoning and proving in mathematics lessons. I gained a lot of confidence in my own competence in the area of reasoning; [the project] afforded a shift from a procedural to a conceptual understanding of R&P.

The third result, for me, is the critical role of students in the process of R&P: "The most important thing I learned during the meeting was probably how easy it can be to 'forget' pupil reasoning in lesson plans";

> Together, these reflections underscore the evolving role of the teacher, not only as a planner of structured lessons but as a facilitator of pupils' reasoning, through the ability to ask the right questions, anticipate difficulties, or continuously assess pupils' understanding.

This issue was addressed in a recent paper studying a Chinese case (Shao et al., 2025).

The final result concerns the collective work of teachers, which, although common in certain countries such as China, is not at all common in the five countries participating in the MaTeK project. This collective work proved to be a valuable resource for the project, both for the work carried out by each national team and, even more so, for the international meeting: "Working in groups suited me especially because we shared different opinions and ideas with each other"; "mainly the collaborative design of lesson plans in mixed groups but also presentations of national lesson plans, received feedback, information about the school system in each country". It was seen as enriching and a significant factor in PMTs' learning experience.

On this issue, I would like to raise two issues: to what extent does teachers' collective work promote the organization of students' collective work in the classroom? How did the differences in curriculum, languages, and cultures between the different teams involved in the project stimulate or, on the contrary, hinder interactions during the international seminar?

9.5 Discussing these results

Having chosen to organize my review and comments chapter by chapter, I have already discussed a number of findings in the previous chapter. In this section, I would like to return to these points in more detail, discussing the points for discussion proposed by the authors.

First, I greatly appreciated the authors' comparison of their findings with those of previous research. And now some remarks.

What is R&P? The fact that, "across groups, during the Intervention (between pre- and post-interviews) many PMTs moved away from viewing R&P solely as the construction of rigorous proofs" appears, for me, as a fundamental result, justifying in itself the Intervention, and it has a very important consequence, due to "the transformative potential of reasoning-and-proving interventions in reshaping teachers' understanding of effective mathematics teaching" (Stylianides, 2009).

What is the place of R&P in mathematics education? Two seemingly contradictory findings emerge: on the one hand, R&P issues cut across all areas of mathematics ("R&P came to be viewed as a fundamental element that cuts across all areas of mathematics"), while on the other hand, certain areas of mathematics seem more conducive to the introduction of R&P elements (often geometry, sometimes algebra). In fact, in my opinion, the contradiction is only apparent, because we need to distinguish between two levels:

- the level of teachers' practice for practising: yes, R&P issues are everywhere, and teachers need to make this clear to their students by giving them opportunities to experience R&P in the different areas of mathematics they explore;
- the level of the curriculum, where the issue is legitimate: in which areas are R&P approaches most necessary? In France, for example, there has been a discussion about the comparative advantages of emphasizing proof issues in geometry or algorithms.

What is the role of teachers? Two results could appear here also as contradictory: on one side, "a stronger focus on the teacher's role in shaping classroom norms that support reasoning and proving"; on the other side, "Teachers positioned as facilitators, guiding pupils through questioning and scaffolding". In my opinion, the contradiction is only apparent; actually, we have to consider these two views as interrelated, and the concept of orchestration (Trouche & Drijvers, 2014) could help to situate the position of teachers in these complex dialectics.

Why is it so difficult, for the PMTs, to embed reasoning within lesson structures? As an explanation, the authors suggest "a need for more explicit scaffolding and professional development to help PMTs transition from theoretical appreciation to practical implementation". This explanation does not entirely convince me. In my view, conceptual integration difficulties cannot simply be resolved by more scaffolding but, on the contrary, by more iteration in the design process. A single implementation in class is not much.... I was struck by how (Trouche, 2024) in China, conceptual deepening involved going back and forth between phases of design, implementation, and redesign, which encouraged participants to think more deeply. Something to consider for a future research project?

How could be explained "the minimal use of digital tools" within the LP? The authors suggest:

> This may be because, in many countries (such as Slovakia, the Czech Republic, and Italy), teachers must reserve a special room for pupils to use digital tools. Consequently, the teacher often uses digital tools for demonstration purposes during mathematics lessons, which is more feasible in the mentioned countries.

This explanation does not quite convince me either. Students now have tools (calculators or tablets) that can be used in regular math classes. And it is a well-established fact that teachers prefer to use technological tools themselves to "show" the representation of mathematical objects because it allows them to maintain control of the teaching situation. If we want them to use these tools to support students' own exploration strategies, we need to make this a necessary component of LP and support PMTs in their design work (e.g., by working on the triplet: mathematical situations – technological tools – R&P). This consideration of technological tools is all the more necessary given that their evolution towards increasingly advanced forms (particularly those incorporating artificial intelligence) will lead to other forms of validation of "truth" in mathematics that may compete with the approaches taken by teachers and students.

9.6 Conclusion, limitations, and implications for practice

I would first like to emphasize my agreement with the main conclusions drawn by the authors, then highlight a few issues that should be explored in greater depth.

The authors highlight four key elements: the centrality of R&P for effective mathematics education; the potential of international interactions for teacher training in an uncertain world; the relevance of a methodology based on "lesson design capacity"; and, finally, the further development of this concept of design capacity, incorporating a range of elements such as sensitivity to linguistic and cultural contexts.

There are numerous implications for the practice of mathematics teacher training, in particular the importance of pedagogical tools (such as teacher moves) that enable teachers to better control their work in the classroom; the need to consider fluid interactions between theory and practice; and the value of international interactions (made easier by remote communication tools).

I would then like to qualify one of the project's conclusions, which I feel is too definitive: "This approach not only ensures the long-term impact of the project but also supports the ongoing preparation of PMTs to meet the demands of diverse and dynamic educational contexts." As the authors themselves note, it would undoubtedly be necessary to monitor PMTs once they enter the profession in order to be able to make such a claim. And it is a recognized fact that it is very difficult for a new teacher to retain all the skills acquired during their training when they join a new school; they will be confronted with more experienced colleagues whose practices are legitimized by their acquired experience. In order to retain the knowledge acquired during training, specific measures are undoubtedly needed: integrating not only teacher trainers but also regular teachers into initial training programmes; linking initial training programmes with continuing education programmes; extending the solidarity built up during the research project by forming small groups of teachers who commit to continuing their interactions for one or two years; and encouraging schools that take on these new teachers to draw on their experience by engaging in a collective LP design process. There are many possible avenues, none of which are obvious, and they must be carefully considered with the people concerned.

Finally, the MaTeK project has begun to explore avenues that need to be pursued: what curriculum should be designed to better take into account R&P processes in a rapidly changing world? Is there an optimal number of iterations (design – implementation – reflection – redesign) to ensure the quality of the resources produced and develop the design capacity of the teachers involved? How can we integrate digital resources (including AI) to benefit R&P processes? And, finally, what is the relationship between experimental mathematics and experimental demonstrations? I always have Lakatos (1976)'s seminal oeuvre in mind. This is a line of thinking that will undoubtedly continue to evolve as mathematics, technology, and teaching methods evolve.

REFERENCES

Adler, J. (2000). Conceptualising resources as a theme for teacher education. *Journal of Mathematics Teacher Education, 3*, 205–224. https://doi.org/10.1023/A:1009903206236

Anichini, G., Arzarello, F., Ciarrapico, L., & Robutti, O. (Eds.) (2004). *Matematica 2003: Attività didattiche e prove di verfica per un nuovo curricolo di matematica (ciclo secondario)*. Matteoni Stampatore.

Anthony, G., & Walshaw, M. (2009). *Effective pedagogy in mathematics: Educational practices series-19*. Gonnet Imprimeur. https://www.iaoed.org/downloads/EdPractices_19.pdf

Arzarello, F., Douek, N., & Falcade, R. (2020). *Mathematical argumentation and proof: Theory, research, and practice in mathematics education*. Springer.

Assis, C., Gitirana, V., & Trouche, L. (2018). The metamorphosis of resource systems of prospective teacher: From studying to teaching. In V. Gitirana, T. Miyakawa, M. Rafalska, S. Soury-Lavergne, & L. Trouche (Eds.), *Proceedings of the Re(s)sources 2018 International Conference* (pp. 39–42). ENS de Lyon

Avşar Tuncay, A. (2022). A study on teacher education in Turkey. In M. S. Khine (Ed.), *Handbook of research on teacher education: Pedagogical innovations and practices in the Middle East*. (pp. 459–477). Springer. https://doi.org/10.1007/978-981-19-2400-2_27

Baccaglini-Frank, A., Antonini, S., Leung, A., & Mariotti, M. A. (2018). From pseudo-objects in dynamic explorations to proof by contradiction. *Digital Experiences in Mathematics Education, 4*, 87–109. https://doi.org/10.1007/s40751-018-0039-2

Backfisch, I., Franke, U., Ohla, K., Scholtz, N., & Lachner, A. (2023). Collaborative design practices in pre-service teacher education for technological pedagogical content knowledge (TPACK): Group composition matters. *Unterrichtswissenschaft, 51*(4), 579–604. https://doi.org/10.1007/s42010-023-00192-z

Balacheff, N. (2010). Bridging knowing and proving in mathematics: An essay from a didactical perspective. In G. Hanna, H. N. Jahnke, & H. Pulte (Eds.), *Explanation and proof in mathematics* (pp. 115–135). Springer.

Ball, D. L., & Forzani, F. (2009). The work of teaching and the challenge for teacher education. *Journal of Teacher Education, 60*(5), 497–511. https://doi.org/10.1177/0022487109348479

Ball, L. D., Thames, M. H., & Phelps, G. (2008). Content knowledge for teaching: What makes it special? *Journal of Teacher Education, 59*(5), 389–407. https://doi.org/10.1177/0022487108324554

Battista, M. T. (2007). The development of geometric and spatial thinking. In F. K. Lester, Jr. (Ed.), *Second handbook of research on mathematics teaching and learning* (pp. 843–908). Information Age Publishing.

Battista, M. T. (2016). Mathematical reasoning and sense making. In M. T. Battista (Ed.), *Reasoning and sense making in the mathematics classroom: Pre-K–Grade 2*. (pp. 1–26). National Council of Teachers of Mathematics.

Bergqvist, T., & Lithner, J. (2012). Mathematical reasoning in teachers' presentations. *The Journal of Mathematical Behavior, 31*(2), 252–269. https://doi.org/10.1016/j.jmathb.2011.12.002

Beswick, K. (2012). Teachers' beliefs about school mathematics and mathematicians' mathematics and their relationship to practice. *Educational Studies in Mathematics, 79*(1), 127–147. https://doi.org/10.1007/s10649-011-9333-2

Bieda, K. (2010). Enacting proof-related tasks in middle school mathematics: Challenges and opportunities. *Journal for Research in Mathematics Education, 41*(4), 351–382. https://doi.org/10.5951/jresematheduc.41.4.0351

Biza, I., Nardi, E., & Zachariades, T. (2009). Teacher beliefs and the didactic contract on visualization. *For the Learning of Mathematics, 29*(3), 31–36. https://flm-journal.org/index.php?do=details&lang=en&vol=29&num=3&pages=31-36

Black, P., & Wiliam, D. (2018). Classroom assessment and pedagogy. *Assessment in Education: Principles, Policy & Practice, 25*(6), 551–575. https://doi.org/10.1080/0969594X.2018.1441807

Boaler, J. (2022). *Mathematical mindsets: Unleashing students' potential through creative math, inspiring messages and innovative teaching*. John Wiley & Sons Inc.

Boero, P. (2007). *Theorems in school: From history, epistemology and cognition to classroom practice*. Sense Publishers

Boesen, J., Helenius, O., Bergqvist, E., Bergqvist, T., Lithner, J., Palm, T., & Palmberg, B. (2014). Developing mathematical competence: From the intended to the enacted curriculum. *The Journal of Mathematical Behavior, 33*, 72–87. https://doi.org/10.1016/j.jmathb.2013.10.001

Boesen, J., Lithner, J., & Palm, T. (2010). The relation between types of assessment tasks and the mathematical reasoning students use. *Educational Studies in Mathematics, 75*, 89–105. https://doi.org/10.1007/s10649-010-9242-9

Boon, M., Orozco, M., & Sivakumar, K. (2022). Epistemological and educational issues in teaching practice-oriented scientific research: Roles for philosophers of science. *European Journal for Philosophy of Science, 12*, Article 16. https://doi.org/10.1007/s13194-022-00447-z

Borko, H., Jacobs, J., Eiteljorg, E., & Pittman, M. (2008). Videotape as a tool for promoting productive discussions in mathematics professional development. *Teaching and Teacher Education, 24*(2), 417–436. https://doi.org/10.1016/j.tate.2006.11.012

Borko, H., Jacobs, J., & Koellner, K. (2010). Creating a framework for teacher professional development: A study of collaborative lesson planning. *Educational Leadership, 67*(6), 58–65. https://doi.org/10.1016/B978-0-08-044894-7.00654-0

Borwein, J., & Devlin, K. (2010). *The computer as crucible: An introduction to experimental mathematics*. A K Peters/CRC Press

Borwein, J. M. (2016). The life of modern homo habilis mathematicus: Experimental computation and visual theorems. In J. Monaghan, L. Trouche, & J. M. Borwein (Eds.), *Tools and mathematics, instruments for learning* (Vol. 110, pp. 23–90). Mathematics Education Library. Springer. https://doi.org/10.1007/978-3-319-02396-0_3

Bowen, G. A. (2009). Document analysis as a qualitative research method. *Qualitative Research Journal, 9*(2), 27–40. https://doi.org/10.3316/QRJ0902027

Brown, M. W. (2009). The teacher-tool relationship: Theorizing the design and use of curriculum materials. In J. T. Remillard, B. A. Herbel-Eisenmann, & G. M. Lloyd (Eds.), *Mathematics teachers at work: Connecting curriculum materials and classroom instruction* (pp. 17–36). Routledge.

Buchbinder, O., & McCrone, S. (2020). Preservice teachers learning to teach proof through classroom implementation: Successes and challenges. *The Journal of Mathematical Behavior, 58*. https://doi.org/10.1016/j.jmathb.2020.100779

Buchbinder, O., & McCrone, S. (2022). Guiding principles for teaching mathematics via reasoning and proving. In J. Hodgen, E. Geraniou, G. Bolondi, & F. Ferretti (Eds.), *Proceedings of the 12th congress of European Research Society in Mathematics Education (CERME12)* (pp. 1–8). Free University of Bozen-Bolzano and ERME. https://hal.archives-ouvertes.fr/hal-03746878

Buchbinder, O., & McCrone, S. (2023). Preparing prospective secondary teachers to teach mathematical reasoning and proof: The case of the role of examples in proving. *ZDM Mathematics Education, 55*, 779–792. https://doi.org/10.1007/s11858-023-01493-4

Buchbinder, O., Weingarden, H., Karsenty, R., & Tabach, M. (2023). Enhancing pre-service teachers' knowledge for teaching proof through a task-based intervention. *ZDM Mathematics Education, 55*(2), 385–399. https://doi.org/10.1007/s11858-022-01397-9

Buchbinder, O., & Zaslavsky, O. (2011). Is this a coincidence? The role of examples in fostering a need for proof. *ZDM Mathematics Education, 43*, 269–281. https://doi.org/10.1007/s11858-011-0324-7

Cai, J. (2025). Improving students' learning of mathematics: Lessons from several decades of research. In X. K. Yan, A. Mamolo, & I. Kontrovich (Eds.), *Where is the mathematics in your math education research? Personal accounts of leading educators*. Springer.

Cai, J., & Cirillo, M. (2014). What do we know about reasoning and proving? Opportunities and missing opportunities from curriculum analyses. *International Journal of Educational Research, 64*, 132–140. https://doi.org/10.1016/j.ijer.2013.10.007

Cevikbas, M., König, J., & Rothland, M. (2024). Empirical research on teacher competence in mathematics lesson planning: Recent developments. *ZDM – Mathematics Education, 56*(1), 101–113. https://doi.org/10.1007/s11858-023-01487-2

Chen, S., & Zhang, B. (2019). Improving prospective teachers' lesson planning knowledge and skills through lesson study. In R. Huang, A. Takahashi, & J. da Ponte (Eds.), *Theory and practice of lesson study in mathematics: Advances in mathematics education* (pp. 549–575). Springer. https://doi.org/10.1007/978-3-030-04031-4_27

Chun, T. C., & Abdullah, M. N. L. Y. (2022). The effects of language teachers' attitudes, barriers and enablers in teaching 21st century skills at Chinese vernacular schools. *Malaysian Journal of Learning and Instruction, 19*(1), 115–146.

Cihan, F., & Akkoç, H. (2019). Developing pre-service mathematics teachers' pedagogical content knowledge of proof schemes: An intervention study. In U. T. Jankvist, M. van den Heuvel-Panhuizen, & M. Veldhuis (Eds.), *Proceedings of the eleventh congress of the European Society for Research in Mathematics Education* (pp. 155–162). Freudenthal Group & Freudenthal Institute, Utrecht University and ERME.

Cirillo, M. (2018). Engaging students with non-routine geometry proof tasks. In P. Herbst, U. Cheah, P. Richard, & K. Jones (Eds.), *International perspectives on the teaching and learning of geometry in secondary schools: ICME-13 monographs*. Springer (pp. 283–300). https://doi.org/10.1007/978-3-319-77476-3_16

Clarà, M., Mauri, T., Colomina, R., & Onrubia, J. (2019). Supporting collaborative reflection in teacher education: A case study. *European Journal of Teacher Education, 42*(2), 175–191. https://doi.org/10.1080/02619768.2019.1576626

Clarke, D., & Hollingsworth, H. (2002). Elaborating a model of teacher professional growth. *Teaching and Teacher Education, 18*(8), 947–967. https://doi.org/10.1016/S0742-051X(02)00053-7

Cohen, L., Manion, L., & Morrison, K. (2007). *Research methods in education* (6th ed.) Routledge.

Coles, A., & Helliwell, T. (2023). The role of mathematics teacher educators in preparing teachers of mathematics to respond to global challenges within their classrooms. *London Review of Education, 21*(1), Article 2. https://doi.org/10.14324/LRE.21.1.02

Conradie, J., & Frith, J. (2000). Comprehension tests in mathematics. *Educational Studies in Mathematics, 42*(3), 225–235. https://doi.org/10.1023/A:1017502919000

Council of Higher Education [CoHE] (2018). *Yeni Öğretmen Yetiştirme Lisans Programları* [New teacher education undergraduate programs]. https://www.yok.gov.tr/en

Corbin, J., & Strauss, A. (2008). *Basics of qualitative research: Techniques and procedures for developing grounded theory* (3rd ed.) Sage.

de Villiers, M. (1990). The role and function of proof in mathematics. *Pythagoras, 24*, 17–24.

de Villiers, M. D. (2012). *Rethinking proof with the geometer's sketchpad*. Key Curriculum Press.

Dewey, J. (1933). *How we think*. Heath and Company.

Dickerson, D. S., & Doerr, H. M. (2014). High school mathematics teachers' perspectives on the purposes of mathematical proof in school mathematics. *Mathematics Education Research Journal, 26*, 711–733. https://doi.org/10.1007/s13394-013-0091-6

Dietiker, L., Singh, R., Riling, M., Nieves, H. I., & Barno, E. (2023). Narrative characteristics of captivating secondary mathematics lessons. *Educational Studies in Mathematics, 112*, 481–504. https://doi.org/10.1007/s10649-022-10184-y

Doerr, H. M., & Lesh, R. (2003). A modeling perspective on teacher development. In R. Lesh, & H. Doerr (Eds.), *Beyond constructivism: Models and modeling perspectives on mathematics problem solving, learning, and teaching*. Lawrence Erlbaum.

Drijvers, P., Stacey, K., & Trouche, L. (2024). Mathematics curricula in the digital era: Policy and implementation. In M. Clements, B. Kaur, T. Lowrie, V. Mesa, & J. Prytz (Eds.), *Fourth international handbook of mathematics education*. Springer Nature. https://doi.org/10.1007/978-3-031-51474-6_28

Durand-Guerrier, V., Boero, P., Douek, N., Epp, S., & Tanguay, D. (2012). Argumentation and proof in the mathematics classroom. In G. Hanna, & M. de Villiers (Eds.), *Proof and proving in mathematics education* (pp. 349–367). Springer.

Durand-Guerrier, V., & Tanguay, D. (2018). Working on proofs as contributing to conceptualization: The case of IR completeness. In A. Stylianides, & G. Harel (Eds.), *Advances in mathematics education research on proof and proving: ICME-13 monographs*. Springer. https://doi.org/10.1007/978-3-319-70996-3_2

Ellis, A. (2011). Generalizing-promoting actions: How classroom collaborations can support students' mathematical generalizations. *Journal for Research in Mathematics Education, 42*(4), 308–345. https://doi.org/10.5951/jresematheduc.42.4.0308

Ellis, A., Özgür, Z., & Reiten, L. (2019). Teacher moves for supporting student reasoning. *Mathematics Education Research Journal, 31*, 107–132. https://doi.org/10.1007/s13394-018-0246-6

English, L. D., & Gainsburg, J. (2016). Problem-solving in a 21st-century mathematics curriculum. In L. D. English, & D. Kirshner (Eds.), *Handbook of international research in mathematics education* (3rd ed., pp. 313–335). Routledge.

Eurydice. (2020). *The Teaching Profession in Europe: Practices, Perceptions, and Policies*. Retrieved February 10, 2025 from https://eurydice.eacea.ec.europa.eu
Evans, T., Mejía-Ramos, J. P., & Inglis, M. (2022). Do mathematicians and undergraduates agree about explanation quality? 445–467. https://doi.org/10.1007/s10649-022-10164-2
Fernández, C., Llinares, S., & Rojas, Y. (2020). Prospective mathematics teachers' development of noticing in an online teacher education program. *ZDM Mathematics Education, 52*, 959–972. https://doi.org/10.1007/s11858-020-01149-7
Fitriati, F., Rosli, R., & Iksan, Z. H. (2023). Enhancing prospective mathematics teachers' lesson planning skills through lesson study within school university partnership program. *Journal on Mathematics Education, 14*(1), 69–84.
Forman, E. A. (2020). Communities of practice in mathematics education. In S. Lerman (Ed.), *Encyclopedia of mathematics education*. Springer. https://doi.org/10.1007/978-3-030-15789-0_25
Forsler, A., Nilsson, P., & Walan, S. (2024). Capturing and developing teachers' pedagogical content knowledge in sustainable development using content representation and video-based reflection. *Research in Science Education, 54*, 393–412. https://doi.org/10.1007/s11165-023-10149-y
Geiger, V., Calder, N., Tan, H., Loong, E., Miller, J., & Larkin, K. (2016). Transformations of teaching and learning through digital technologies. In K. Makar, S. Dole, J. Visnovska, M. Goos, A. Bennison, & K. Fry (Eds.), *Research in mathematics education in Australasia 2012–2015*. Springer. https://doi.org/10.1007/978-981-10-1419-2_13
Goldsmith, L. T., Doerr, H. M., & Lewis, C. C. (2014). Mathematics teachers' learning: A conceptual framework and synthesis of research. *Journal of Mathematics Teacher Education, 17*, 5–36. https://doi.org/10.1007/s10857-013-9245-4
Gravemeijer, K., & Doorman, M. (1999). Context problems in realistic mathematics education: A calculus course as an example. *Educational Studies in Mathematics, 39*, 111–129. https://doi.org/10.1023/A:1003749919816
Gravemeijer, K., & Prediger, S. (2019). Topic-specific design research: An introduction. In G. Kaiser, & N. Presmeg (Eds.), *Compendium for early career researchers in mathematics education* (pp. 33–57). Springer Open. https://doi.org/10.1007/978-3-030-15636-7_2
Grossman, P., Wineburg, S., & Woolworth, S. (2001). Toward a theory of teacher community. *Teachers College Record, 103*(6), 942–1012. https://doi.org/10.1111/0161-4681.00140
Guskey, T. R. (2002). Professional development and teacher change. *Teachers and Teaching: Theory and Practice, 8*(3), 381–391. https://doi.org/10.1080/135406002100000512
Hanna, G. (1990). Some pedagogical aspects of proof. *Interchange, 21* (1), 6–13. https://doi.org/10.1007/BF01809605
Hanna, G. (2000). Proof, explanation, and exploration: An overview. *Educational Studies in Mathematics, 44*(1–2), 5–23. https://doi.org/10.1023/A:1012737223465
Hanna, G., & Barbeau, E. (2010). Proofs as bearers of mathematical knowledge. *ZDM Mathematics Education, 40*, 345–353. https://doi.org/10.1007/s11858-008-0080-5
Hanna, G., & de Villiers, M. (2012). Aspects of Proof in Mathematics Education. In G. Hanna, & M. de Villiers (Eds.), *Proof and proving in mathematics education: New ICMI study series* (Vol. 15, pp. 1–10). Springer. https://doi.org/10.1007/978-94-007-2129-6_1
Hansen, A., Mavrikis, M., & Geraniou, E. (2016). Supporting teachers' technological pedagogical content knowledge of fractions through co-designing a virtual manipulative. *Journal of Mathematics Teacher Education, 19*, 205–226. https://doi.org/10.1007/s10857-016-9344-0

Harel, G., & Rabin, J. M. (2010). Teaching practices that can promote the authoritative proof scheme. *Canadian Journal of Science, Mathematics and Technology Education*, *10*(2), 139–159. https://doi.org/10.1080/14926151003778282

Harel, G., & Sowder, L. (1998). Students' proof schemes: Results from exploratory studies. *American Mathematical Society*, *7*, 234–283.

Harel, G., & Sowder, L. (2007). Toward comprehensive perspectives on the learning and teaching of proof. In F. K. Lester (Ed.), *Second handbook of research on mathematics teaching and learning* (pp. 805–842). Information Age.

Hattie, J., & Timperley, H. (2007). The power of feedback. *Review of Educational Research*, *77*(1), 81–112. https://doi.org/10.3102/003465430298487

Heinze, A., & Reiss, K. (2007). Reasoning and proof in the mathematics classroom. *Analysis*, *27*(2–3), 333–357. https://doi.org/10.1524/anly.2007.27.2-3.333

Herbert, S., Vale, C., Bragg, L. A., Loong, E., & Widjaja, W. (2015). A framework for primary teachers' perceptions of mathematical reasoning. *International Journal of Educational Research*, *74*(2015), 26–37. https://doi.org/10.1016/j.ijer.2015.09.005

Herbst, P., & Kosko, K. (2014). Mathematical knowledge for teaching and its specificity to high school geometry instruction. In J. J. Lo, K. Leatham, & L. Van Zoest (Eds.), *Research trends in mathematics teacher education: Research in mathematics education*. Springer. https://doi.org/10.1007/978-3-319-02562-9_2

Hiebert, J., Carpenter, T. P., Fennema, E., Fuson, K. C., Wearne, D., Murray, H., Olivier, A., & Human, P. (1997). *Making sense: Teaching and learning mathematics with understanding*. Heinemann.

Hiim, H., & Hippe, E. (2006). *Praksisveiledning i lærerutdanningen. En didaktisk veiledningsstrategi* [*Practice guidance in teacher training. A didactic guidance strategy*]. Gyldendal akademisk.

Hogan, T., Rabinowitz, M., & Craven, J. A. (2003). Problem representation in teaching: Inferences from research of expert and novice teachers. *Educational Psychologist*, *38*(4), 235–247. https://doi.org/10.1207/S15326985EP3804_3

Hoyles, C., & Noss, R. (2019). The meaning of statistical variation in the context of work. In R. A. Lesh, E. Hamilton, & J. J. Kaput (Eds.), *Foundations for the future in mathematics education*. Routledge.

Huang, H.-M. E., & Witz, K. G. (2011). Developing children's conceptual understanding of area measurement: A curriculum and teaching experiment. *Learning and Instruction*, *21*(1), 1–13. https://doi.org/10.1016/j.learninstruc.2009.09.002

Hußner, I., Lazarides, R., Symes, W., Richter, E., & Westphal, A. (2023). Reflect on your teaching experience: Systematic reflection of teaching behaviour and changes in student teachers' self-efficacy for reflection. *Zeitschrift für Erziehungswissenschaft*, *26*, 1301–1320. https://doi.org/10.1007/s11618-023-01190-8

Inglis, M., & Alcock, L. (2012). Expert and Novice Approaches to Reading Mathematical Proofs. *Journal for Research in Mathematics Education*, *43*(4), 358–390. https://doi.org/10.5951/jresematheduc.43.4.0358

INVALSI. (2018). *Il quadro di riferimento delle prove di matematica del sistema nazionale di valutazione*. INVALSI.

Jacobs, V. R., & Spangler, D. A. (2017). Research on core practices in K–12 mathematics teaching. In J. Cai (Ed.), *Compendium for research in mathematics education* (pp. 766–792). National Council of Teachers of Mathematics.

Jaworski, B. (2006). Theory and practice in mathematics teaching development: Critical inquiry as a mode of learning in teaching. *Journal of Mathematics Teacher Education*, *9*, 187–211. https://doi.org/10.1007/s10857-005-1223-z

Jaworski, B., & Huang, R. (2014). Teachers and didacticians: Key stakeholders in the processes of developing mathematics teaching. *ZDM Mathematics Education*, *46*, 173–188. https://doi.org/10.1007/s11858-014-0574-2

Jeannotte, D., & Kieran, C. A. (2017). Conceptual model of mathematical reasoning for school mathematics. *Educational Studies in Mathematics, 96*, 1–16. https://doi.org/10.1007/s10649-017-9761-8

Johnson, R., Severance, S., Penuel, W. R., & Leary, H. (2015). Teachers, tasks, and tensions: Lessons from a research-practice partnership. *Journal of Mathematics Teacher Education, 19*, 169–185. https://doi.org/10.1007/s10857-015-9338-3

Jones, K. (2010), The role of the teacher in teaching proof and proving in geometry, *Proceedings of the BritishSociety for Research into Learning Mathematics, 30*(2), 62–67

Jones, K., & Pepin, B. (2016). Research on mathematics teachers as partners in task design. *Journal of Mathematics Teacher Education, 19*, 105–121. https://doi.org/10.1007/s10857-016-9345-z

Karsenty, R., Dole, S., Clivaz, S., Griese, B., & Pöhler, B. (2024). Roles, identities and interactions of various participants in mathematics teacher collaboration. In H. Borko, & D. Potari (Eds), *Teachers of mathematics working and learning in collaborative groups*. New ICMI Study Series. Springer. https://doi.org/10.1007/978-3-031-56488-8_4

Kieran, C., Doorman, M., & Ohtani, M. (2015). Frameworks and principles for task design. In A. Watson, & M. Ohtani (Eds.), *Task design in mathematics education*. New ICMI Study Series. Springer. https://doi.org/10.1007/978-3-319-09629-2_2

Kilpatrick, J., Swafford, J., & Findell, B. (Eds.) (2001). *Adding it up: Helping children learn mathematics*. The National Academies Press.

Knuth, E. J. (2002). Secondary school mathematics teachers' conceptions of proof. *Journal for Research in Mathematics Education, 33*(5), 379–405. https://doi.org/10.2307/4149959

Ko, Y. Y. (2010). Pre-service mathematics teachers' conceptions of proof: Implications for educational research. *International Journal of Science and Mathematics Education, 8*, 1109–1129. https://doi.org/10.1007/s10763-010-9235-2

Kordaki, M., & Potari, D. (2002). The effect of area measurement tools on student strategies: The role of a computer microworld. *International Journal of Computers for Mathematical Learning, 7*, 65–100. https://doi.org/10.1023/A:1016051411284

Korthagen, F. A. J. (2004). In search of the essence of a good teacher: Towards a more holistic approach to teacher education. *Teaching and Teacher Education, 20*(1), 77–97. https://doi.org/10.1016/j.tate.2003.10.002

Lakatos, I. (1976). *Proofs and refutations: The logic of mathematical discovery*. Cambridge University Press.

Lave, J., & Wenger, E. (1991). *Situated learning: Legitimate peripheral participation*. Cambridge University Press.

Lesseig, K. (2016). Investigating mathematical knowledge for teaching proof in professional development. *International Journal of Research in Education and Science, 2*(2), 253–270. Retrieved May 2025 from https://files.eric.ed.gov/fulltext/EJ1105101.pdf

Lesseig, K., Hine, G., & Boardman, K. (2018). Preservice secondary mathematics teachers' perceptions of proof in the secondary mathematics classroom. In T. E. Hodges, G. J. Roy, & A. M. Tyminski (Eds.), *Proceedings of the 40th annual meeting of the North American chapter of the international group for the psychology of mathematics education*. University of South Carolina & Clemson University.

Lester, F.K., & Cai, J. (2016). Can mathematical problem solving be taught? Preliminary answers from 30 years of research. In: P. Felmer, E. Pehkonen, J. Kilpatrick (Eds.), *Posing and solving mathematical problems. Research in mathematics education*. Cham: Springer. https://doi.org/10.1007/978-3-319-28023-3_8

Lewis, C. C., Perry, R. R., Friedkin, R., & Roth, J. R. (2012). Improving teaching does improve teachers: Evidence from lesson study. *Journal of Teacher Education, 63*(5), 368–375. https://doi.org/10.1177/0022487112446633

Liljedahl, P. (2015). Numeracy task design: A case of changing mathematics teaching practice. *ZDM Mathematics Education, 47*, 625–637. https://doi.org/10.1007/s11858-015-0703-6

Lim, W., Son, J. W., & Kim, D. J. (2018). Understanding preservice teacher skills to construct lesson plans. *International Journal of Science and Mathematics Education, 16*, 519–538. https://doi.org/10.1007/s10763-016-9783-1

Lithner, J. A. (2008). Research framework for creative and imitative reasoning. *Educational Studies in Mathematics, 67*, 255–276. https://doi.org/10.1007/s10649-007-9104-2

Mariotti, M. A. (2006). Proof and proving in mathematics education. In Á. Gutiérrez & P. Boero (Eds.), *Handbook of research on the psychology of mathematics education* (pp. 173–204). Brill.

Mariotti, M. A., Durand-Guerrier, V., & Stylianides, G. J. (2018). Argumentation and proof. In T. Dreyfus, M. Artigue, D. Potari, S. Prediger, & K. Ruthven (Eds.), *Developing research in mathematics education* (pp. 75–89). Routledge.

Mason, J. (2011). Noticing: Roots and branches. In M. G. Sherin, V. R. Jacobs, & R. A. Philipp (Eds.), *Mathematics teacher noticing: Seeing through teachers' eyes* (pp. 35–50). Routledge.

Michal, J. (2023). Pre-service teachers' conceptions and beliefs of reasoning and proof in school mathematics. In P. Drijvers, C. Csapodi, H. Palmér, K. Gosztonyi, & E. Kónya (Eds.), *Proceedings of the Thirteenth Congress of the European Society for Research in Mathematics Education (CERME13)* (pp. 192–199). Alfréd Rényi Institute of Mathematics; ERME.

Ministero dell'Istruzione, dell'Università e della Ricerca [MIUR] (2010). Schema di regolamento recante "Indicazioni nazionali riguardanti gli obiettivi specifici di apprendimento concernenti le attività e gli insegnamenti compresi nei piani degli studi previsti per i percorsi liceali". https://www.indire.it/lucabas/lkmw_file/licei2010/indicazio-ni_nuovo_impaginato/_decreto_indicazioni_nazionali.pdf

Ministero dell'Istruzione dell'Università e della Ricerca [MIUR] (2012). *Indicazioni nazionali per il curricolo della scuola dell'infanzia e del primo ciclo d'istruzione. Annali della pubblica Istruzione, Periodico multimediale per la scuola italiana a cura del Ministero dell'Istruzione, dell'Università e della Ricerca*. MIUR

Ministero dell'Istruzione e del Merito [MIUR] (2025). Nuove indicazioni per la scuola dell'infanzia e primo ciclo di istruzione 2025. Retrieved March 16, 2025 from https://www.mim.gov.it/documents/20182/0/Nuove+indicazioni+2025.pdf/cebce5de-1e1d-12de-8252-79758c00a50b?version=1.1&t=1741806754378

Ministry of Education and Research (2019). *Curriculum for mathematics year 1–10.* https://data.udir.no/kl06/v201906/laereplaner-lk20/MAT01-05.pdf?lang=eng

Ministry of Education, Research, Development and Youth of the Slovak Republic [MinEdu] (2023). Štátny vzdelávací program pre základné vzdelávanie. *Vzdelávacie štandardy: Matematika a Informatika* [State curriculum for primary and lower secondary education. Educational Standards: Mathematics and informatics]. Retrieved January 12, 2025 from https://www.minedu.sk/data/files/11820_marematika-a-informatika.pdf

Ministry of National Education [MoNE] (2018a). Matematik dersi öğretim programı (İlkokul ve ortaokul 1, 2, 3, 4, 5, 6, 7 ve 8. sınıflar). [Mathematics curriculum (Elementary and middle school grades 1, 2, 3, 4, 5, 6, 7 and 8)].

Ministry of National Education [MoNE] (2018b). Ortaöğretim matematik dersi öğretim programı (9, 10, 11, 12. sınıflar). [Secondary mathematics curriculum (Grades 9, 10, 11, 12)].

Ministry of National Education [MoNE] (2024a). Ortaokul matematik dersi öğretim programı (5, 6, 7 ve 8. Sınıflar) [Mathematics curriculum (Middle school grades 5, 6, 7 and 8)], Türkiye yüzyılı maarif modeli.

Ministry of National Education [MoNE] (2024b). Ortaöğretim matematik dersi öğretim programı (Hazırlık, 9, 10, 11 ve 12. Sınıflar) [Secondary mathematics curriculum (Preparatory, Grades 9, 10, 11, and 12), Türkiye yüzyılı maarif modeli].
Mitchell, R. N., & Marin, K. A. (2015). Examining the use of a structured analysis framework to support prospective teacher noticing. *Journal of Mathematics Teacher Education, 18*, 551–575.
Morris, A. K. (2007). Factors affecting pre-service teachers' evaluation of the validity of students' mathematical arguments in classroom contexts. *Cognition and Instruction, 25*(4), 479–522. https://doi.org/10.1080/07370000701632405
MŠMT (2023). Rámcový vzdělávací program pro základní vzdělávání [Framework educational programme for basic education]. https://www.edu.cz/wp-content/uploads/2021/07/RVP-ZV-2021.pdf
Nardi, E., & Knuth, E. (2017). Changing classroom culture, curricula, and instruction for proof and proving: How amenable to scaling up, practicable for curricular integration, and capable of producing long-lasting effects are current interventions? *Educational Studies in Mathematics, 96*(2), 267–274.
National Council of Teachers of Mathematics [NCTM] (2000). *Principles and standards for school mathematics*. National Council of Teachers of Mathematics [NCTM].
National Council of Teachers of Mathematics [NCTM] (2009). *Focus in high school mathematics: Reasoning and sense making*. National Council of Teachers of Mathematics [NCTM].
Niss, M., & Højgaard, T. (2019). Mathematical competencies revisited. *Educational Studies in Mathematics, 102*(1), 9–28. https://doi.org/10.1007/s10649-019-09903-9
OECD (2018). *PISA for development assessment and analytical framework: Reading, mathematics and science*. OECD.
OECD (2022). PISA 2022 mathematics framework. OECD. https://pisa2022-maths.oecd.org/ca/index.html
OECD (2023). *The future of education and skills. Learning compass for mathematics*. https://www.oecd.org/content/dam/oecd/en/about/projects/edu/education-2040/publications/OECD-Learning-Compass-for-Mathematics-2023-13-Oct.pdf
Orr, S., & Bieda, K. (2023). Learning to elicit student thinking: The role of planning to support academically rigorous questioning sequences during instruction. *Journal of Mathematics Teacher Education*. https://doi.org/10.1007/s10857-023-09603-5
Otten, S., Candela, A. G., de Araujo, Z., Haines, C., & Munter, C. (2019). *Proceedings of the forty-first annual meeting of the North American chapter of the international group for the psychology of mathematics education*. University of Missouri.
Pedemonte, B. (2007). How can the relationship between argumentation and proof be analyzed? *Educational Studies in Mathematics, 66*(1), 23–41. https://doi.org/10.1007/s10649-006-9057-x
Pepin, B. (2015). *Enhancing mathematics/STEM education: A "resourceful" approach*. Inaugural Lecture. Technische Universiteit Eindhoven. https://pure.tue.nl/ws/files/8804776/Pepin2015.pdf
Pepin, B., Artigue, M., Gitirana, V., Miyakawa, T., Ruthven, K., & Xu, B. (2019). Mathematics teachers as curriculum designers: An international perspective to develop a deeper understanding of the concept. In L. Trouche, G. Gueudet, & B. Pepin (Eds.), *The 'resource' approach to mathematics education: Advances in mathematics education* (pp 121–143). Springer. https://doi.org/10.1007/978-3-030-20393-1_6
Pepin, B., Gueudet, G., & Trouche, L. (2013). Re-sourcing teachers' work and interactions: A collective perspective on resources, their use and transformations. *ZDM - Mathematics Education, 45*(7), 929–944. https://doi.org/10.1007/s11858-013-0534-2

Pepin, B., Gueudet, G., & Trouche, L. (2017). Refining *teacher design capacity*: Mathematics teachers' interactions with digital curriculum resources. *ZDM Mathematics Education*, *49*(5), 799–812. https://rdcu.be/tmXb

Pepin, B., Kock, Z. J., & Rezat, S. (2023). Toward student agency in the selection and use of digital resources for learning and studying mathematics. In B. Pepin, G. Gueudet, & J. Choppin (Eds.), *Handbook of digital resources in mathematics education: Springer International Handbooks of Education*. Springer. https://doi.org/10.1007/978-3-030-95060-6_26-1

Perry, S., & Lewis, C. (2009). The Role of Lesson Planning in Improving Teacher Quality. *Journal of Teacher Education*, *60*(5), 515–530

Philipp, A. (2007). Mathematics teachers' beliefs and affect. In F. K. Lester (Ed.), *Second handbook of research on mathematics teaching and learning*. National Council of Teachers of Mathematics.

Potari, D., & da Ponte, J. P. (2017). Current research on prospective secondary mathematics teachers' knowledge. In *The mathematics education of prospective secondary teachers around the world: ICME-13 topical surveys*. Springer. https://doi.org/10.1007/978-3-319-38965-3_2

Psycharis, G., Potari, D., & Krog Skott, C. (2023). Addressing collective and individual aspects of teacher design with digital resources in collaborative settings. In B. Pepin, G. Gueudet, & J. Choppin (Eds.), *Handbook of digital resources in mathematics education: Springer International Handbooks of Education*. Springer. https://doi.org/10.1007/978-3-030-95060-6_26-1

Reid, D. A., & Knipping, C. (2010). *Proof in mathematics education: Research, learning, and teaching*. Brill.

Remillard, J., & Heck, D. J. (2014). Conceptualising the curriculum enactment process in mathematics education. *ZDM Mathematics Education*, *46*(5), 705–718. https://doi.org/10.1007/s11858-014-0600-4

Remillard, J. T. (2005). Examining key concepts in research on teachers' use of mathematics curricula. *Review of Educational Research*, *75*(2), 211–246. https://doi.org/10.3102/00346543075002211

Roller, S. A. (2016). What they notice in video: A study of prospective secondary mathematics teachers learning to teach. *Journal of Mathematics Teacher Education*, *19*, 477–498. https://doi.org/10.1007/s10857-015-9307-x

Saldaña, J. (2015). *The coding manual for qualitative researchers*. Sage.

Sánchez-Matamoros, G., Fernández, C., & Llinares, S. (2019). Relationships among prospective secondary mathematics teachers' skills of attending, interpreting and responding to students' understanding. *Educational Studies in Mathematics*, *100*, 83–99. https://doi.org/10.1007/s10649-018-9855-y

Santagata, R., & Guarino, J. (2011). Using video to teach future teachers to learn from teaching. *ZDM Mathematics Education*, *43*, 133–145. https://doi.org/10.1007/s11858-010-0292-3

Schoenfeld, A. H. (2010). *How we think: A theory of goal-oriented decision making and its educational applications* (1st ed.) Routledge. https://doi.org/10.4324/9780203843000

Schoenfeld, A. H. (2011). Noticing matters. A lot. Now what? In M. G. Sherin, V. R. Jacobs, & R. A. Philipp (Eds.), *Mathematics teacher noticing: Seeing through teachers' eyes* (pp. 223–238). Routledge.

Schön, D. A. (1983). *The reflective practitioner: How professionals think in action*. Routledge.

Sevinc, S., Kohanová, I., Isiksal-Bostan, M., Kubáček, Z., Isler-Baykal, I., Lada, M., Çakıroğlu, E., & Di Paola, B. (2022). Developing an integrated framework for analyzing ways of reasoning in mathematics. In *ICERI2022 Proceedings* (pp. 2082–2089). Spain.

Shao, M., Trgalova, J., & Trouche, L. (2025, online). Teaching 3D geometry with dynamic environments in upper secondary classes: How to help students combine empirical perception and theoretical deduction? *Journal of Mathematics Teacher Education*. https://doi.org/10.1007/s10857-025-09716-z

Sherin, M., Jacobs, V., & Philipp, R. (Eds.) (2010). *Mathematics teacher noticing: Seeing through teachers' eyes* (1st ed.). Routledge.

Shulman, L. S. (1986). Those who understand: Knowledge growth in teaching. *Educational Researcher*, *15*(2), 4–14. https://doi.org/10.3102/0013189X015002004

Shute, V. J. (2008). Focus on formative feedback. *Review of Educational Research*, *78*(1), 153–189. https://doi.org/10.3102/0034654307313795

Silverman, B., & Even, R. (2015). Textbook explanations: Modes of reasoning in 7th-grade Israeli mathematics textbooks. In *Proceedings of CERME 9* (pp. 205–212). https://hal.archives-ouvertes.fr/hal-01281094/document

Smith, M. S., & Stein, M. K. (2011). *5 practices for orchestrating productive mathematical discussions*. Reston: National Council of Teachers of Mathematics.

Smith, M. S., & Stein, M. K. (2018). *Five practices for orchestrating productive mathematics discussions*. Heinemann.

Stacey, K., & Vincent, J. (2009). Modes of reasoning in explanations in Australian eighth-grade mathematics textbooks. *Educational Studies in Mathematics*, *72*(3), 271–288. https://doi.org/10.1007/s10649-009-9193-1

Star, J. R., & Strickland, S. K. (2008). Learning to observe: Using video to improve preservice mathematics teachers' ability to notice. *Journal of Mathematics Teacher Education*, *11*, 107–125. https://doi.org/10.1007/s10857-007-9063-7

State Pedagogical Institute [SPI] (2014a). Štátny vzdelávací program: Matematika – nižšie stredné vzdelávanie [State educational program: Mathematics - lower secondary education]. https://www.statpedu.sk/files/articles/dokumenty/inovovany-statny-vzdelavaci-program/matematika_nsv_2014.pdf

State Pedagogical Institute [SPI] (2014b). Štátny vzdelávací program: Matematika – gymnázium so štvorročným a päťročným vzdelávacím programom [State educational program: Mathematics – grammar school with a four-year and five-year educational program.online]. https://www.statpedu.sk/files/articles/dokumenty/inovovany-statny-vzdelavaci-program/matematika_g_4_5_r.pdf

State Pedagogical Institute [SPI] (2015a). *Štátny vzdelávací program: nižšie stredné vzdelávanie – 2. stupeň základnej školy* [State educational program: lower secondary education – 2nd level of primary school]. https://www.statpedu.sk/files/articles/dokumenty/inovovany-statny-vzdelavaci-program/svp_nsv_6_2_2015.pdf

State Pedagogical Institute [SPI] (2015b). Štátny vzdelávací program pre gymnáziá (úplné stredné všeobecné vzdelávanie) [State educational program for gymnasiums (complete secondary general education)]. https://www.statpedu.sk/files/articles/dokumenty/inovovany-statny-vzdelavaci-program/statny_vzdel_program_pre_gymnazia.pdf

Stein, M. K., Engle, R. A., Smith, M. S., & Hughes, C. (2008). Orchestrating productive discussions: Norms for engaging students in mathematical thinking. *Mathematics Teaching in the Middle School*, *13*(4), 160–167.

Stein, M. K., Smith, M. S., Henningsen, M., & Silver, E. A. (2009). *Implementing standards-based mathematics instruction: A casebook for professional development*. Teachers College Press.

Stephan, M., & Akyuz, D. (2012). A proposed instructional theory for integer addition and subtraction. *Journal for Research in Mathematics Education*, 43(4), 428–464. https://doi.org/10.5951/jresematheduc.43.4.0428

Stephan, M. (2020). Sociomathematical norms in mathematics education. In S. Lerman (Ed.) *Encyclopedia of mathematics education*. Springer. https://doi.org/10.1007/978-3-030-15789-0_143

Strachota, S., Knuth, E., & Blanton, M. (2018). Cycles of generalizing activities in the classroom. In: C. Kieran (Ed.) *Teaching and learning algebraic thinking with 5- to 12-year-olds*. ICME-13 Monographs. Springer. https://doi.org/10.1007/978-3-319-68351-5_15

Stylianides, A. J. (2007). Proof and proving in school mathematics. *Journal for Research in Mathematics Education, 38*(3), 289–321. https://www.jstor.org/stable/30034869

Stylianides, A. J. (2009). Reasoning-and-proving in school mathematics textbooks. *Mathematical Thinking and Learning, 11*(4), 258–288. https://doi.org/10.1080/10986060903253954

Stylianides, A. J. (2019). Secondary students' proof constructions in mathematics: The role of written versus oral mode of argument representation. *Review of Education, 7*(1), 156–182. https://doi.org/10.1002/rev3.3157

Stylianides, A. J., & Ball, D. L. (2008). Understanding and describing mathematical knowledge for teaching: knowledge about proof for engaging students in the activity of proving. *Journal of Mathematics Teacher Education, 11*, 307–332. https://doi.org/10.1007/s10857-008-9077-9

Stylianides, A. J., & Stylianides, G. J. (2013). Seeking research-grounded solutions to problems of practice: classroom-based interventions in mathematics education. *ZDM Mathematics Education, 45*(3), 333–341. https://doi.org/10.1007/s11858-013-0501-y

Stylianides, A. J., & Stylianides, G. J. (2022). Introducing students and prospective teachers to the notion of proof in mathematics. *The Journal of Mathematical Behavior, 66*. https://doi.org/10.1016/j.jmathb.2022.100957

Stylianides, G., Stylianides, A., & Weber, K. (2017). Research on the teaching and learning of proof: Taking stock and moving forward. In J. Cai (Ed.) *Compendium for research in mathematics education*. National Council of Teachers of Mathematics.

Stylianides, G. J. (2008). An analytic framework of reasoning-and-proving. *For the Learning of Mathematics, 28*(1), 9–16. https://flm-journal.org/Articles/308086F06226BBFBA6966CF21B6EC.pdf

Stylianides, G. J., & Stylianides, A. J. (2009). Facilitating the transition from empirical arguments to proof. *Journal for Research in Mathematics Education, 40*(3), 314–352. https://doi.org/10.5951/jresematheduc.40.3.0314

Stylianides, G. J., & Stylianides, A. J. (2017). Research-based interventions in the area of proof: The past, the present, and the future. *Educational Studies in Mathematics, 96*(2), 119–127. https://doi.org/10.1007/s10649-017-9782-3

Stylianides, G. J., Stylianides, A. J., & Moutsios-Rentzos, A. (2024). Proof and proving in school and university mathematics education research: A systematic review. *ZDM Mathematics Education, 56*, 47–59. https://doi.org/10.1007/s11858-023-01518-y

Tataroğlu Taşdan, B., Tekin Dede, A., & Yiğit Koyunkaya, M. (2022). Examining pre-service mathematics teachers' argumentation-supported lesson plans and their noticing during planning. *International Journal of Mathematical Education in Science and Technology, 55*(6), 1309–1329. https://doi.org/10.1080/0020739X.2022.2054741

Tempier, F. (2015). New perspectives for didactical engineering: An example for the development of a resource for teaching decimal number system. *Journal of Mathematics Teacher Education, 19*, 261–276. https://doi.org/10.1007/s10857-015-9333-8

Thanheiser, E., Olanoff, D., Hillen, A., Feldman, Z., Tobias, J. M., & Welder, R. M. (2015). Reflective analysis as a tool for task redesign: The case of prospective elementary teachers solving and posing fraction comparison problems. *Journal of Mathematics Teacher Education, 19*, 123–148. https://doi.org/10.1007/s10857-015-9334-7

The NRICH Primary Team. (2014, July). *Reasoning: Identifying opportunities (Article)*. University of Cambridge, NRICH. https://nrich.maths.org/10990

Thomas, J., Jong, C., Fisher, M. H., & Schack, E. O. (2017). Noticing and knowledge: Exploring theoretical connections between professional noticing and mathematical knowledge for teaching. *The Mathematics Educator, 26*(2), 3–25. https://doi.org/10.63301/tme.v26i2.2030

Thompson, A. G. (1992). Teachers' beliefs and conceptions: A synthesis of the research. In D. A. Grouws (Ed.), *Handbook of research on mathematics teaching and learning*. Macmillan.

Trgalová, J., & Tabach, M. (2024). Pre-service teachers' development of digital resource design capacity. *ZDM Mathematics Education, 56*, 651–665. https://doi.org/10.1007/s11858-024-01554-2

Trouche, L. (2016). Integrating tools as an ordinary component of the curriculum in mathematics education. In J. Monaghan, L. Trouche, & J. M. Borwein, *Tools and mathematics, instruments for learning* (pp. 267–304). Springer.

Trouche, L. (2024). Chine 2024: Les contradictions d'un système d'enseignement en mutation. *Au Fil des Maths revue de l'APMEP, 554*, 57–67

Trouche, L. (2025). A key element in curriculum developments around the world: Teachers' resources. *International Education Congress*, Ankara, 3 & 4 June 2025.

Trouche, L., Adler, J., & Remillard, J. T. (2023). Conceptualizing teachers' interactions with resources in crossing languages and cultures. *ZDM Mathematics Education, 55*, 497–519. https://doi.org/10.1007/s11858-023-01488-1

Trouche, L., & Drijvers, P. (2014). Webbing and orchestration. Two interrelated views on digital tools in mathematics education. *Teaching Mathematics and Its Applications: International Journal of the Institute of Mathematics and Its Applications, 33*(3), 193–209. https://doi.org/10.1093/teamat/hru014

Trouche, L., Gueudet, G., & Pepin, B. (Eds.) (2019). *The 'resource' approach to mathematics education*. Springer Series: Advances in Mathematics Education. Springer. https://doi.org/10.1007/978-3-030-20393-1

Trouche, L., Gueudet, G., & Pepin, B. (2020). Documentational approach to didactics. In S. Lerman (Ed.), *Encyclopedia of mathematics education* (2nd ed., pp. 237–247). Springer. https://doi.org/10.1007/978-3-319-77487-9_100011-1

Ulusoy, F., & İncikabi, L. (2023). Preservice mathematics teachers' selection of curriculum resources in individual and group lesson planning processes. *International Journal of Mathematical Education in Science and Technology, 54*(4), 557–578. https://doi.org/10.1080/0020739X.2021.1958944

Unione Matematica Italiana. [UMI]. (2013). *Proposta di un Syllabus di matematica per i Licei Scientifici (nuovo ordinamento)* [Proposal for a mathematics syllabus for scientific high schools (new system)]. https://umi.dm.unibo.it/wp-content/uploads/2013/10/Syllabus-UMI-V-liceo-2014.pdf

Utdanningsdirektoratet (2020). *The Norwegian National Curriculum (LK20) – Mathematics*. Oslo: Norwegian Directorate for Education and Training.

Vale, I., & Barbosa, A. (2024). Exploring the creative potential of mathematical tasks in teacher education. *International Electronic Journal of Mathematics Education, 19*(4), em0790. https://doi.org/10.29333/iejme/15075

Valenta, A., & Enge, O. (2020). Bevisrelaterte kompetanser i læreplanen LK20 for matematikk i grunnskolen [Proof-related competences in the LK20 curriculum for mathematics in primary school]. *Acta Didactica Norden, 14*(3). https://doi.org/10.5617/adno.8195

Valenta, A., Rø, K., & Klock, S. I. (2024). A framework for reasoning in school mathematics: analyzing the development of mathematical claims. *Educational Studies in Mathematics, 116*, 91–111. https://doi.org/10.1007/s10649-024-10309-5

van den Akker, J. (2004). Curriculum perspectives: An introduction. In J. van den Akker, W. Kuiper, & U. Hameyer (Eds.), *Curriculum landscapes and trends* (pp. 1–10). Springer. https://doi.org/10.1007/978-94-017-1205-7_1

van den Akker, J. (2013). Curricular development research as a specimen of education design research. In T. Plomp, & N. Nieveen (Eds.), *Educational design research* (pp. 52–71), Netherlands Institute for Curriculum Development.

van Es, E. A., & Sherin, M. G. (2002). Learning to notice: Scaffolding new teachers' interpretations of classroom interactions. *Journal of Technology and Teacher Education*, *10*(4), 571–596.

van Es, E. A., & Sherin, M. G. (2008). Mathematics teachers' "Learning to Notice" in the context of a video club. *Teaching and Teacher Education*, *24*, 244–276. https://doi.org/10.1016/j.tate.2006.11.005

van Es, E. A., & Sherin, M. G. (2024). Correction: Expanding on prior conceptualizations of teacher noticing. *ZDM Mathematics Education*. https://doi.org/10.1007/s11858-024-01570-2

Vondrová, N., Novotná, M., Pavlasová, L., Robová, J., Stará, J., & Uličná, K. (2020). *Video-interventions: Bridges between theory and practice in pre-service teachers' development*. Karolinum Press.

Waite, S. (2010). Teaching and learning outside the classroom: Personal values, alternative pedagogies and standards. *Education 3–13*, *39*(1), 65–82. https://doi.org/10.1080/03004270903206141

Ward, J. R., & McCotter, S. S. (2004). Reflection as a visible outcome for preservice teachers. *Teaching and Teacher Education*, *20*(3), 243–257. https://doi.org/10.1016/j.tate.2004.02.004

Wartofsky, M. W. (1973). Perception, representation, and the forms of action: Towards a historical epistemology. In M. Wartofsky (Ed.), *Models: Representation and the scientific understanding*. Reidel.

Watson, A., & Ohtani, M. (2015). *Task design in mathematics education: The 22nd ICMI study*. Springer.

Watson, A., & Sullivan, P. A. (2008). Teachers learning about tasks and lessons. In D. Tirosh, & T. Wood (Eds.), *International handbook of mathematics teacher education: Tools and processes in mathematics teacher education* (1st ed., pp. 109–134, Vol. 2). Sense Publishers.

Weingarden, M., & Buchbinder, O. (2023). Teacher learning to teach mathematics via reasoning and proving: A discursive analysis of lesson plans modifications. *Frontiers in Education*, *8*. https://doi.org/10.3389/feduc.2023.1154531

Wenger, E. (1998). *Communities of practice: Learning, meaning, and identity*. Cambridge University Press. https://doi.org/10.1017/CBO9780511803932

Wenger, E., McDermott, R. A., & Snyder, W. (2002). *Cultivating communities of practice: A guide to managing knowledge*. Harvard Business Press.

Whitcomb, J. A. (2003). Learning and pedagogy in initial teacher preparation. In I. B. Weiner (Ed.) *Handbook of psychology* (Vol. 7, pp. 533–556). Wiley.

Williams, N. N., Williams, B. K., Jones-Fosu, S., & Carter, T. (2023). An examination of cross-cultural experiences on developing culturally responsive teacher candidates. *Emerald Open Research*, *1*(3). https://doi.org/10.1108/EOR-03-2023-0004

Xu, S., & Connelly, M. (2022). Reciprocal learning in teacher education between Canada and China in a globalized context. In *The palgrave handbook of teacher education research*. Palgrave Macmillan. https://doi.org/10.1007/978-3-030-59533-3_37-1

Yackel, E., & Cobb, P. (1996). Sociomathematical norms, argumentation, and autonomy in mathematics. *Journal for Research in Mathematics Education*, *27*(4), 458–477. https://doi.org/10.2307/749877

Yenmez, A. A., & Celik, T. (2021). The study of knowledge exchange on sociomathematical norms: Dialogue writing and dramatization. *Journal of Teacher Education and Educators, 10*(3), 369–394. https://dergipark.org.tr/en/pub/jtee/issue/67769/908138

Yin, R. K. (2009). *Case study research: Design and methods.* (4th ed.) Sage.

Yopp, D. A. (2009). From inductive reasoning to proof. *Mathematics Teaching in the Middle School, 15*(5), 286–291. National Council of Teachers of Mathematics.

Yunianto, W., Prahmana, R. C. I., & Crisan, C. (2021). Indonesian mathematics teachers' knowledge of content and students of area and perimeter of rectangle. *Journal on Mathematics Education, 12*(2), 223–238. https://doi.org/10.22342/jme.12.2.13537.223-238

Zaslavsky, O. (2019). There is more to examples than meets the eye: Thinking with and through mathematical examples in different settings. *The Journal of Mathematical Behavior, 53*, 245–255. https://doi.org/10.1016/j.jmathb.2017.10.001

Zaslavsky, O., Chapman, O., & Leikin, R. (2011). Professional development of mathematics educators: Trends and tasks. In *Second international handbook of mathematics education* (pp. 877–911). https://doi.org/10.1007/978-94-010-0273-8_28

Zazkis, R., & Leikin, R. (2010). Advanced mathematical knowledge in teaching practice: Perceptions of secondary mathematics teachers. *Mathematical Thinking and Learning, 12*(4), 263–281. https://doi.org/10.1080/10986061003786349

Zhu, J., Huang, X., & Trouche, L. (2025,). The development of didactic praxeologies in lesson design through social resource interactions: Evidence from two novice mathematics teachers. *Educational Studies in Mathematics, 119*, 487–513. https://doi.org/10.1007/s10649-025-10403-2

APPENDIX 1

Example of assignment (Assignment 1)

Reading part:

Use guidance for reading the paper:

Silverman, B., & Even, R. (2015). Textbook explanations: Modes of reasoning in 7th-grade Israeli mathematics textbooks. In *Proceedings of CERME 9* (pp. 205–212). Charles University, Faculty of Education, and ERME. https://hal.archives-ouvertes.fr/hal-01281094/document

Practical part:

1 Select a task which supports/leads to reasoning and proof (R&P) for grade 8 or 9.
2 Why do you think this task addresses R&P? (Why do you choose this task?) Explain.
3 Find out where R&P is mentioned in the curriculum. What does the curriculum say about R&P?
4 Identify the type(s) of reasoning that is/are required in your task and relate your task to modes of reasoning in Silverman and Even's (2015) paper.
5 Pupils discussed the following statement: *"Every odd whole number is the sum of two consecutive whole numbers."*

   a Write an argument showing that this statement is true. What mode of reasoning did you apply?
   b Construct a few possible arguments that might be expected from pupils when discussing the validity of this statement.

APPENDIX 2

The final assignment

Individual reflective text

Write a reflective text (2–4 pages) on your work over the three/four sessions. Here are some prompts for you to consider. Elaborate on them and explain your reasoning.

1. What have you learned?
2. What has surprised you?
3. What have you found beneficial?
4. How did the work in the group help you?
5. How confident do you now feel that you will be able to plan a mathematics lesson towards a given goal?
6. How confident do you now feel that you will be able to support your pupils' R&P?
7. What do you still need to work on in terms of developing your pupils' R&P?

APPENDIX 3

Individually assigned questionnaire after the international meeting of PMTs in Prague

- Which part of the programme in Prague did you appreciate the most? Explain why.
- What was the most important thing(s) you learned during the Prague meeting? Explain why.
- In which ways/how did the Prague meeting help you to enhance your own thinking on lesson development on reasoning and proof? Explain why.
- Describe your experience of working in the international group on Saturday.
- Do you have any other comments related to the Prague meeting? Please write them here.

APPENDIX 4

Exit cards used

Date:

Student name:

1 Select the smiley that best describes your overall feeling about your work today.

Explain why you selected this smiley:

2 Describe the most important thing(s) you learned today.

3 Which part of the session did you like best today? Explain why. / Describe the activity you liked the best today and explain why.

4 Describe what you found difficult today, and explain why.

5 Select the smiley that best describes how confident you feel about designing a lesson focused on R&P.

Explain why you selected this smiley:

APPENDIX 5

Part 3 from the Czech lesson plan worksheet

PART 3 – "Write an algebraic expression to the given picture and draw a picture to the given algebraic expression."

A math worksheet with geometric shapes labeled with variables (a, b) and numbers. Problems involve writing expressions for areas of shapes and drawing shapes for given expressions.

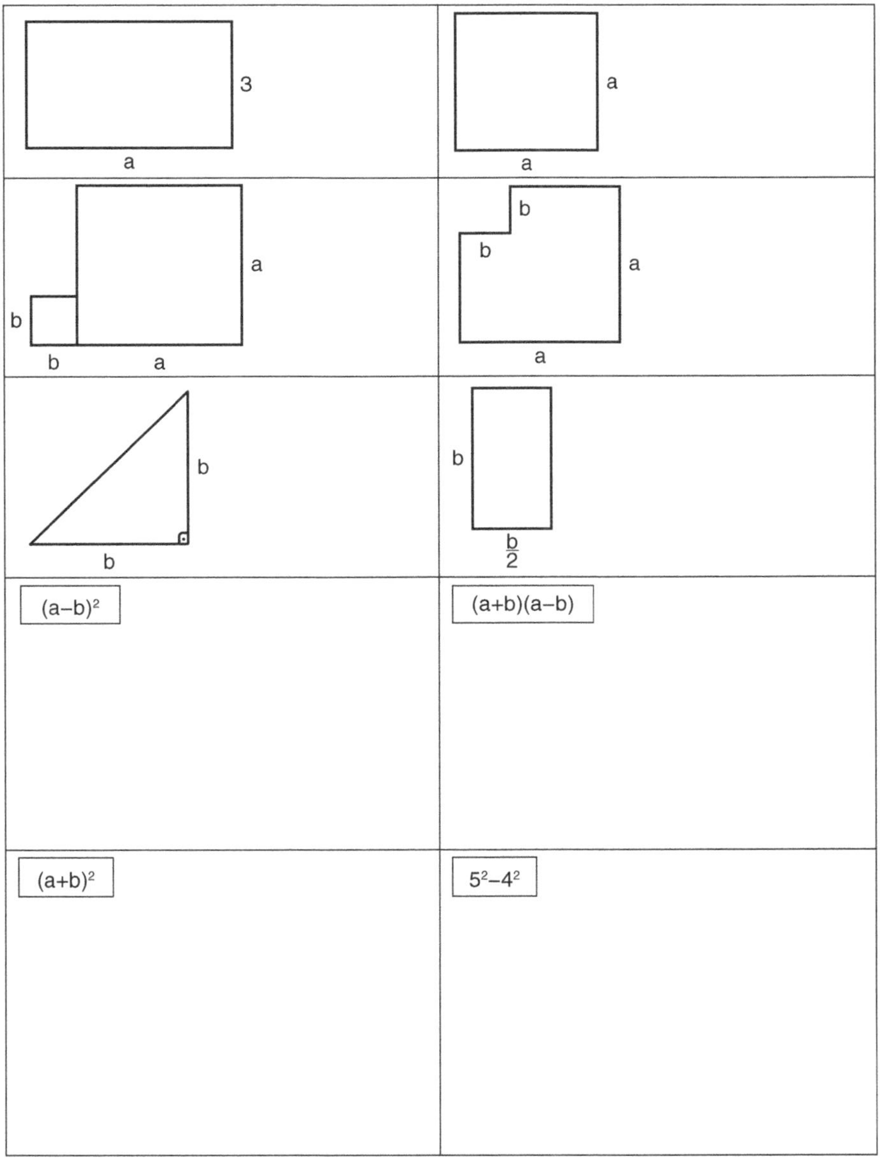

INDEX

Pages in *italics* refer to figures, pages in **bold** refer to tables, and pages followed by "n" refer to notes.

For Product Safety Concerns and Information please contact our EU representative GPSR@taylorandfrancis.com Taylor & Francis Verlag GmbH, Kaufingerstraße 24, 80331 München, Germany